PLANT DEVELOPMENT

T. Pullaiah
K.C. Naidu
K. Lakshminarayana
B. Hanumantha Rao

2024
Regency Publications
A Division of
Astral International Pvt. Ltd.
New Delhi – 110 002

First Published, 2007
Reprinted, 2024

ISBN: 9788189233587 (HB)

Publisher's Note:

Every possible effort has been made to ensure that the information contained in this book is accurate at the time of going to press, and the publisher and author cannot accept responsibility for any errors or omissions, however caused. No responsibility for loss or damage occasioned to any person acting, or refraining from action, as a result of the material in this publication can be accepted by the editor, the publisher or the author. The Publisher is not associated with any product or vendor mentioned in the book. The contents of this work are intended to further general scientific research, understanding and discussion only. Readers should consult with a specialist where appropriate.

Every effort has been made to trace the owners of copyright material used in this book, if any. The author and the publisher will be grateful for any omission brought to their notice for acknowledgement in the future editions of the book.

Published by : **Regency Publications**
A Division of
Astral International Pvt. Ltd.
– ISO 9001:2015 Certified Company
4736/23, Ansari Road, Darya Ganj
New Delhi-110 002
Ph. 011-43549197, 23278134
E-mail: info@astralint.com
Website: www.astralint.com

CONTENTS

CHAPTER I

SEED GERMINATION AND SEEDLING GROWTH

Metabolism of Nucleic Acids, Proteins and Mobilization of Food Reserves: Tropisms; Hormonal Control of Seedling Growth; Gene Expression

INTRODUCTION

Seed is the representative of next generation of plant. The function of the seed is to establish a new plant. To fulfill its function, seed has to germinate to give rise to the seedling and eventually matures into an adult plant. Germination commences with the uptake of water by the dry seed (imbibition). The process involves considerable amount of metabolic activities. The enzymes necessary for the initiation of metabolic activity are present within the dry seed. Resumption of respiratory activity commences. Then events including the synthesis of nucleic acids, proteins, mobilization of the storage reserves are associated with growth of the seedling and continued until the seedling becomes autotropic.

METABOLISM OF NUCLEIC ACIDS

During germination and early seedling growth, both cell division and cell elongation take place. At the same time both proteins synthesis and nucleic acids synthesis take place. Later, various substances including nuclei acids move out of the cotyledons and are transferred to other parts of the growing embryo. It decreases in their amount in the storage tissue and increases in the growing embryo. As a result of it, an increase in the weight of the hypocotyls and loss in weight in the cotyledons has been observed. In this process although both DNA and RNA increase in amount, they do not increase at the same rate.

DNA synthesis occurs before or after cell division, at the beginning or at the end of the cell cycle. There is a change in the amount of phosphorylated protein in the nucleus. Some such proteins disappear and others are synthesized. Protein kinases seem to have a regulatory role at some stage of the cell cycle. The amount of polysomes is very low or entirely absent in dry seeds. Studies on cotton seed revealed that ribosome formation and RNA synthesis occurs during the early stages of germination. The cotton seeds also contain long-lived mRNA and the *de novo* synthesis of a protease. Another thing is that the mRNA was formed during embryogenesis, but its translation is blocked at this stage. In some cases ribosomes must be synthesized, while in other cases sufficient ribosomes are present in the embryo for polysome formation.

The long-lived mRNA may not be present in all seeds. Such mRNA is seen in the ribosomal fraction of dry seed embryos. A part of the long-lived mRNA lack polyA and during germination a new polyA RNA is formed. In addition to long-lived mRNA, new mRNA is formed rapidly after the onset of germination. During germination, in addition to mRNA formation, other forms of RNA, namely r-RNA and t-RNA are formed.

In addition to the presence of long-lived mRNA, the dry seeds such as peas, rye, wheat etc. contain ribonucleoproteins, which are capable of sustaining protein synthesis activity. Only a few of the enzymes involved in RNA metabolism have been studied. RNA polymerase activity increases prior to RNA synthesis and the enzymes required for RNA synthesis are present in the mature wheat seed.

PROTEINS AND THEIR MOBILIZATION

Storage Proteins: Plants accumulate protein reserves in developing seeds. The major amount of these reserves consists of specific storage proteins like globulins, which predominate in dicotyledonous seeds, or prolamins, which are the major storage proteins of most cereals. Proteins are generated mainly during seed production and stored either in the cotyledons or in the endosperm, which serve as nitrogen sources for the developing embryo during germination. They are deposited in membrane-bounded compartments, the protein bodies (PB) with diameter from 0.1 to 25 μm, and are generated directly from the endoplasmic reticulum (ER), as in maize or from the protein storage vacuoles (PSV) as in majority of species. The average protein content of cereal grains is 10–15% of their dry weight and that of Leguminosae seeds 20–25%.

The proteins can show the following properties.

- The proteins have no enzymatic activities.
- They serve as nitrogen source for the germinating seeds.

- They occur normally in an aggregated stage within a membrane. (Protein bodies, aleuron grains).
- They are often built from a number of different polypeptide chains.

Protein bodies may contain only one type of storage protein. In certain legumes, for example, some protein bodies contain only albumin or vicilin or legumins, while others contain vicilin and legumin.

According to Osborne's classification, seed proteins can be divided into four classes on the basis of their solubility. (i) albumins — soluble in water and dilute buffers at neutral pHs, (ii) globulins — soluble in salt solutions, (iii) glutelins — soluble in dilute acid or alkali solutions, (iv) prolamins — soluble in aqueous alcohols (70–90%). Characteristically, storage proteins are oligomeric, and are made up of two or more subunits. These subunits are made up of a number of polypeptide chains. The number of genes encoding the different proteins varies considerably among the different prolamins.

Legume seeds are the second most important protein sources, on a world basis, after cereals. The major storage proteins are globulins, which account for up to 70% of the total seed nitrogen. The globulins consists of two major families of proteins which differ in molecular mass and which sediment during ultracentrifugation with sedimentation coefficients (S values) of approximately 7 (average 7–8) and 11 (average 11–13). These are the 7 S (called the vicilin group) and 11 S (the legumin group) proteins.

Hydrolysis of storage protein (polypeptides) to their constituent amino acids requires a class of enzymes called proteinases, some of which effect total hydrolysis where as others produce small polypeptides that must be degraded further by peptidases. Cotyledons and endosperm of dry and germinating seeds contain a variety of proteolytic enzymes some of which are present already in the dry seeds while others appear during germination. The proteolytic enzymes can be divided into *proteinases* and *peptidases* depending on the size of the molecule. The peptidases can be divided into *endopeptidases* and the *exopeptidases*. Exopeptidases can be divided into *carboxypeptidases* and *aminopeptidases*. The liberated amino acids are utilized for protein synthesis in other tissues.

Protein Mobilization: The mobilization of storage proteins is one of the most important post-germinate events in the development of seedling. During germination period, the storage proteins are degraded by a variety of proteases, which convert the insoluble storage proteins into soluble peptides and free amino acids. These are mobilized to the embryonic axis to support and also provide energy by oxidation of the carbon skeleton after deamination. The protein reserve mobilization can be studied in both cereals and dicots.

In Cereals: In cereals, the proteases responsible for the degradation of storage proteins are synthesized in aleurone layer and the synthesis of the enzymes is upregulated by gibberellins and down regulated by ABA (Abscisic acid). Reserve proteins are stored in two regions of the cereal grain: (a) in the aleurone grains of the aleurone layer, and (b) in the protein bodies of the starchy endosperm. The role of the embryo or embryonic axis in mobilization of storage proteins is minor and less understood. Proteinases are produced in the aleurone layer, in some species under the control of gibberellin, and these may be synthesized *de novo*. One or more of these enzymes hydrolyze the proteins within the aleurone grains, and the resulting amino acids are utilized for the synthesis of proteins (e.g., α-amylase) in this tissue.

Proteinases, effective in breaking down the protein reserves within the starchy endosperm, may come from two sources: the aleurone layer or the starchy endosperm itself. Endopeptidases that are released from the aleurone layer may play a central role in endosperm protein hydrolysis in barley. The preformed proteinases that are present within the endosperm of dry grains may be involved in protein mobilization. In addition to their action in degrading reserve proteins, proteinases in the starchy endosperm activate and release enzymes to dissolute the cell walls by hydrolyzing links between glucan and protein components.

Proteolytic activity within the starchy endosperm results in the production of amino acids, dipeptides, and a number of small oligopeptides. The embryo, rapidly absorbs these soluble products via the scutellum, for its growth.

In dicots: Storage protein mobilization has been predominantly investigated in seeds of cultivated plants such as grain legumes, rape, sunflower etc., that during domestication, underwent selection for increased protein accumulation in cotyledon and other storage tissue. In cotyledons storage protein mobilization only becomes measurable after the radicle has protruded and germination is finished. At this stage, inactive precursors of proteinases are synthesized on the rER, from where they are transported through the endomembrane system into PB. Here enzyme activation takes place and protein breakdown is initiated.

The first group of proteinases to increase in activity following germination, as a result of *de novo* synthesis, have been called "proteinase(s) A", which are usually SH-dependent acid endopeptidases. These enzymes act on the insoluble native 11 S (legumin) and 7 S (vicilin) proteins, resulting in the following consecutive series of events.

a) Short-chain peptides are cleaved from the native proteins, which results in a marked increase in their susceptibility to proteinases.
b) Further proteolysis leads to an increase in their negative charge, perhaps resulting from the removal of short basic peptides.
c) An increased solubilization of the proteins resulting from the initial proteolysis.

d) A stepwise decrease in the size of the acidic subunits of the 11 S proteins that are cleaved into polypeptides of intermediate size (10–18 kDa, from the native 20–40 kDa).

The native storage protein, here a di-sulphide linked insoluble legumin, is initially trimmed by proteinase A (endopeptidase) activity (A) to render it more soluble as small oligopeptides are released. Further proteinase A activity and that of protein B endopeptidases (A and B) result in hydrolysis of protein to amino acids and small peptides which are transported into cytosol. Hydrolysis of peptides in the protein bodies is aided by carboxypeptidases (C). The released oligopeptides are degraded further by aminopeptidases and peptidases with in the cytosol to yield amino acids.

The cellular changes that accompany proteolysis have been studied most thoroughly in the cotyledons of mung beans (*Vigna radiata*), in which the major storage protein is vicilin that comprises some 70–80% of the total, and the enzyme responsible for its hydrolysis is an endopeptidase, vicilin peptidohydrolase; with some participation by a carboxypeptidase. The ER is the site of vicilin peptidohydrolase synthesis, and associated with the synthesis of this enzyme the ER undergoes various modifications. Initially the vicilin is cleaved from 50-to 63-kDa components to 20–30-kDa components, and these are then hydrolyzed more slowly.

Then the protein bodies come to contain hydrolytic enzymes other than the peptidohydrolase. During protein hydrolysis a *de novo*-synthesized ribonuclease is inserted into the protein body (after synthesis on ER and transport via vesicles). These organelles have other hydrolase activities also: α-mannosidase and a glucosaminidase (for the hydrolase of mannose and glucosamine residues from vicilin, a glycoprotein), acid phosphatase, phosphodiesterase, and phospholipase D (for digestion of membrane phospholipids).

As protein digestion proceeds, the emptying storage vesicles fuse to form a large vacuole containing hydrolytic enzymes, analogous in function to the lysosome. Digestion of cell contents by the enzymes of the vacuole is achieved when an autophagic process internalizes vesicles. Therefore, disintegration of the cotyledons is achieved by cellular autolysis.

MOBILIZATION OF FOOD RESERVES

The mobilization of stored reserves in the storage organs commences after radicle elongation. Then the reserve food that is present within the storage organs of the seed are mobilized and are converted into soluble forms and transported to the sites where they are required (growing organs) for the support of energy. During germination, the enzymes hydrolyze the stored products such as carbohydrates, lipids, phosphates, proteins etc.

Stored Carbohydrate Catabolism

The most common reserve carbohydrate in seeds is starch. Other stored carbohydrates like hemicelluloses (galactomannans), are to be found as cell wall components of the endosperm of legumes. There are two catabolic pathways of starch: one hydrolytic and the other phosphorolytic.

The amylose and amylopectin in the native starch grain are first hydrolyzed by α-amylase, which breaks the α(1→4) glycosidic links between the glucose residues. The released oligosaccharides are further hydrolyzed by α-amylase until glucose and maltose are produced:

$$\text{Amylose} \xrightarrow{\alpha\text{-Amylase}} \text{Glucose + maltose}$$

$$\text{Amylopectin} \xrightarrow{\alpha\text{-Amylase}} \text{Glucose + maltose + limit dextrin}$$

$$\underset{\text{(Limit dextrinase)}}{\text{Limit dextrin}} \xrightarrow{\substack{\text{Isoamylase}\\ \text{Debranching}\\ \text{(R-) enzyme}}} \underset{\text{oligomers}}{\text{Glucose}} \xrightarrow{\alpha\text{-Amylase}} \text{Maltose + glucose} \xrightarrow{\alpha\text{-Glucosidase}} \text{Glucose}$$

The β-amylase cannot hydrolyze native starch grains; rather it cleaves away successive maltose units. Amylopectin also cannot be completely hydrolyzed but yields maltose and limit dextrin. Further maltose is hydrolyzed into glucose by maltase.

$$\text{Amylose} \xrightarrow{\beta\text{-Amylase}} \text{Maltose}$$

$$\text{Amylopectin} \xrightarrow{\beta\text{-Amylase}} \text{Maltose + limit dextrin}$$

$$\text{Maltose} \xrightarrow[\text{(Maltase)}]{\alpha\text{-Glucosidase}} \text{Glucose}$$

Mobilization of Stored Carbohydrate Reserves in Cereals

It is of two types — the embryo reserves and the endosperm reserves.

Embryo reserves: Low-molecular-weight sugars are stored in lower quantities within the embryo. These probably provide an early source of food substrate during germination and early seedling growth, until the stored food materials available from the endosperm. Sucrose and raffinose are present in the ratio of about 3:1 in dry seeds of barley, and both decline their levels after imbibition.

Endosperm reserves: The starch in the endosperm is degraded to glucose by the action of four enzymes: α-amylase, β-amylase, the debranching enzyme(s), and α-glucosidase. While mobilizing starch and carbohydrates, cell walls of endosperm are also hydrolyzed by these enzymes.

The α-amylase occurs in the region of the scutellum and aluerone layers. In barley, the α-amylase is released into the starchy endosperm from the scutellum. Later the enzyme is also synthesized within the aleurone layer and it hydrolyses the starch reserves. Although α-amylase from the scutellum is important at early stages of mobilization, in barley, maize, wheat etc., later most hydrolysis is affected by isoenzymes from the aleurone layer.

β-Amylase is involved in the digestion of cereal starch. This enzyme is not *de novo* synthesized. Also, it is not released from either the scutellum or the aleurone layer. But it is present in the starchy endosperm of the mature, dry and ungerminated grains. It becomes active after initial digestion of the starch grains by α-amylase from scutellum and aleurone layer.

Amylolysis of the amylose component of starch can be achieved by α- and β-amylases, but digestion of amylopectin ceases with the production of limit dextrins. For further degradation, debranching enzymes are needed probably the limit dextrinase.

The major product of α- and β-amylolysis is maltose. That is hydrolyzed by α-glucosidase (maltase), an enzyme present in low amounts in the embryo and aleurone layer of ungerminated barley. The enzyme increases in activity at both the sites after germination. Hydrolysis of the endosperm cell walls accompanies the digestion of the starch reserves.

MOBILIZATION OF STORED FOOD RESERVES IN DICOTS

Large amounts of proteins, lipids, and starch are accumulated in seeds of dicots during seed development and maturation. In some plants, hemicelluloses, especially galactomannans and mannans are deposited in the cell walls of storage organs. These reserves are mobilized, and the products are used as building blocks for macromolecules and as fuel for energy during seedling growth.

Lipids (triacylglycerols) are hydrolyzed by lipases to yield fatty acids and glycerol. Fatty acids are used for membrane synthesis and for gluconeogenesis *via* β oxidation and the glyoxylate cycle (Fig. 1.1).

Several proteases (e.g., cysteine endopeptidases, aspartic proteases, other endo- and exopeptidases) and α-amylases have also been isolated from a number of plants. Some of these enzymes are stored in the dry seed whereas others are synthesized *de novo* from amino acids released by the hydrolysis of stored proteins.

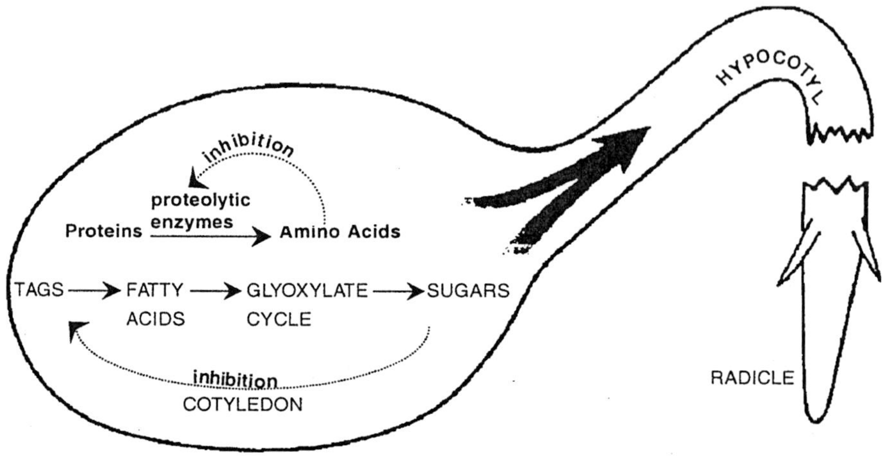

Figure 1.1: Mobilization of reserve proteins and triacylglycerols (TAGs) in cotyledons of cucumber (after Bewley and Black, 1994).

The digestion of galactomannans in cell walls requires the action of three enzymes: an endo- β-mannanase, α-galactosidase and β-mannosidase (Fig. 1.2). It is believed that GAS and/or cytokinins from the embryonic axis move to the endosperm where they bring about the *de novo* synthesis of

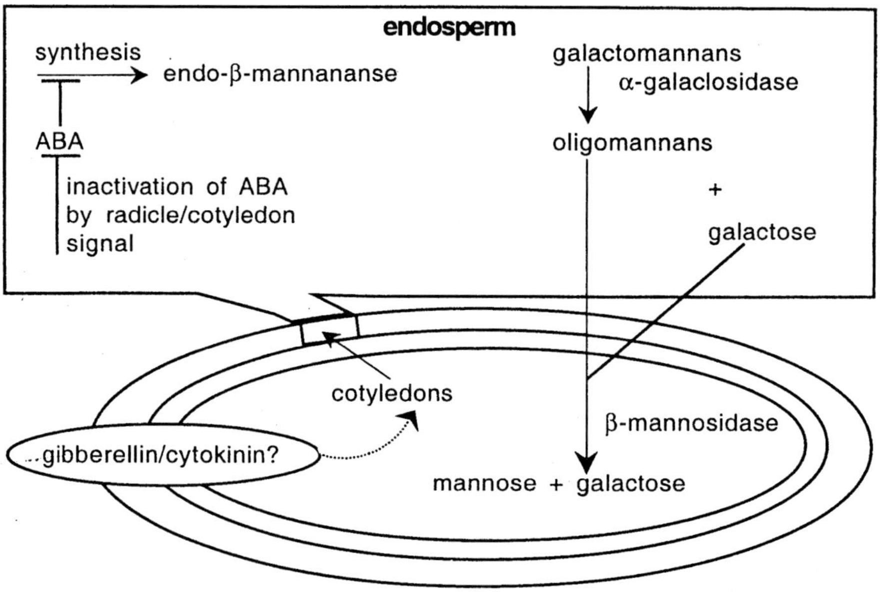

Figure 1.2: Hydrolysis of galactomannans in endosperm cell walls during or immediately following germination (after Bewley, 1997).

endo- β-mannanase and α-galactosidase. The oligomannans and galactose residues are released by the action of these two enzymes. Then they move from the endosperm into the cotyledons where the oligomannans are further broken down into mannose residues by the enzyme β-mannosidase. The products of hydrolysis, galactose and mannose residues are converted into sucrose and utilized for seedling growth.

In legumes, where galactomannan deposits in cell walls of endosperm are so extensive that the cells die (e.g., fenugreek). An aleurone-like layer surrounding the dead tissue synthesizes and secretes the three wall hydrolases. In other endospermic legumes (e.g., carob), where the deposits may not be so heavy, the endosperm cells are metabolically active after germination and are the source of hydrolases.

Some hormone from the embryonic axis moves to the cotyledons or from the embryo to the endosperm and triggers the activation or *de novo* synthesis of hydrolases.

Stored Triacylglycerol Catabolism

Triacylglycerols are the storage lipids in seeds. Their initial hydrolysis (lipolysis) is by lipases, enzymes that catalyze the three-stage hydrolytic cleavage of the fatty acid ester bonds in triacylglycerols (TAGs) to yield glycerol and free fatty acids (FFA).

$$\text{Triacylglycerol} \longrightarrow \text{Diacylglycerol} + \text{FFA} \longrightarrow \text{Monoacylglycerol} + 2\text{FFA} \longrightarrow \text{Glycerol} + 3\text{FFA}$$

Glycerol enters the glycolytic pathway after its phosphorylation and oxidation to the triose phosphates (dihydroxyacetone phosphate — glyceraldehyde-3-phosphate), which are then condensed by aldolase in the reversal of glycolysis to yield hexose units. Alternatively, the triose phosphates may be converted to pyruvate and then oxidized through the citric acid cycle.

The FFA released by lipase may be utilized in oxidation reactions to yield compounds containing fewer carbon atoms. The predominant oxidation pathway is β-oxidation (and glyoxylate cycle).

Catabolism of tricylglycerol reserves within the storage tissues of germinated seeds involves three distinct organelles found within oil-containing cells. These are, (a) the TAG-storing oil body (hydrolysis of TAGs to FFA and glycerol commences within the oil bodies), (b) the glyoxysome (within the glyoxysome the fatty acids are oxidized and synthesis of succinate occurs via the glyoxylate cycle), and (c) the mitochondrion (the succinate is converted to malate or oxaloacetate within the mitochondrian). The malate or oxaloacetate is processed further in the cytosol to give rise sucrose.

Utilization of the Liberated Amino Acids

The amides i.e., asparagines and glutamine are the forms of amino acids transferred to the growing organs of seedling. Later, these amino acids are further metabolized.

$$\text{Aspartate} + \text{glutamine} + \text{ATP} \xrightarrow{\text{Asparagine synthetase}} \text{Asparagine} + \text{glutamate} + \text{AMP} + PPi$$

Glutamine itself is formed from glutamate as follows:

$$\text{Glutamate} + NH_3 + \text{ATP} \xrightarrow{\text{Glutamine synthetase}} \text{Glutamine} + \text{ADP} + Pi$$

$$\text{Glutamine} + \alpha\text{-ketoglutarate} + \text{NAD (P) H} \xrightarrow{\text{GOGAT}} 2\ \text{Glutamate} + \text{NAD (P)}$$

The amino acids that are released from the storage protein by hydrolysis in Castor are aspartate, glutamate, alanine, glycine, and serine and are converted to sucrose and transported as the sugar.

TROPISMS

Plant responses to environmental stimuli including unilateral light, gravity and water, which involve altered growth, are known as Tropisms. They may be positive involving growth towards the stimulus, or negative, away from the stimulus.

Tropisms involve a receptor to sense the stimulus involved and a mechanism to transmit that stimulus to the region of the plant, in which the altered growth will take place.

The different tropisms are,

Phototropism : Orientation towards/away from a unilateral light.
Geotropism : Growth towards/away from gravity.
Thigmotropism : Growth towards touch (a tendril curling around a support).
Hydrotropism : Growth towards water.
Chemotropism : Growth towards a chemical stimulus.

Roots are positively gravitropic and non-phototropic. Shoots are positively phototropic and negatively gravitropic.

Tropisms have attracted a lot of interest since Charles Darwin first studied them in the 1870s. Much of the work has concentrated on the responses of the coleoptile of grasses. The coleoptile is a sheath that covers and protects the primary leaf, and shows rapid growth.

Photoreception

When a coleoptile is exposed to unilateral light for a short period (1–2 hrs) it bends towards the light. The phototropically active light has to be absorbed by blue light, β-carotene and riboflavin. Among various wave lengths of lights, the blue light is the most effective in causing phototropic curvature. In a seedling, the photoreceptor pigment may be located in coleoptile tip.

Mechanism

Auxin was thought to be transmitting signal in phototropism. The bending of the shoot or root is due to differential growth on the two sides of unilaterally exposed organs. In shoot tissues, the growth on the darkened or shaded side is more than on the illuminated side. In the phototropic responsive roots, the situation is reverse; the growth rate is retarded on the shaded side. The unequal distribution of auxins was thought to result from a combination of three different mechanisms. (i) Photodestruction of auxins on the illuminated side. Auxin causes increased growth on the dark side of the shoot or coleoptile stimulating acidification of the cell wall, resulting in loosening of the wall matrix. Turgor pressure against the loosened wall results in elongation.

Geotropism (Gravitropism)

The perception of gravity requires the presence of an object that responds and shows movement to the gravitational field. Such objects are called statoliths. Starch-filled plastids (amyloplasts) move against endoplasmic reticulum (ER) at the cortex of specialized cells termed statocytes (Figure 1.3). The statocytes are in the root cap or in a layer, the starch sheath, adjacent to the vascular tissue. Removal of the root cap results in a loss of sensitivity to gravity in primary roots. The *scr* (*scarecrow*) mutant of *Arabidopsis*, which has no layer of starch-containing cells around its vascular tissue, does not respond to gravity.

Mechanism

Auxin is thought to transmit the signal from the statocytes to the growth zone. In roots, high levels of auxin on the lower side inhibit growth, while lower levels on the upper side permit it; in shoots, it is vice-versa. The transport and unequal distribution of auxin is caused by the changing permeability of cells due to the migration of amyloplasts. Calcium has also been suggested to be involved in graviresponse.

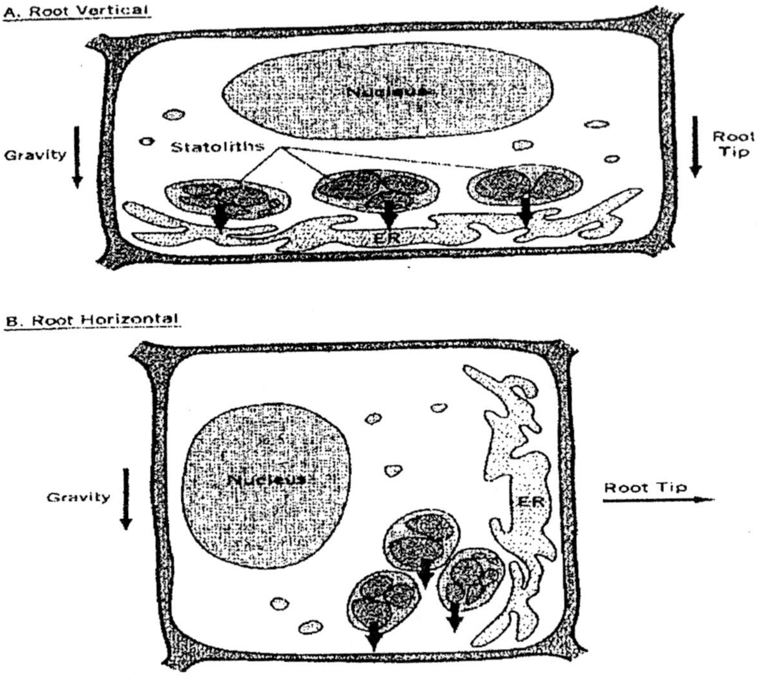

Figure 1.3: Schematic representation of the graviperception model. (A) When the root is oriented vertically, starch-laden statoliths exert pressure uniformly on the endoplasmic reticulum; (B) In the horizontal position, the statoliths are redistributed in the cell in response to gravity, exerting unequal pressure on the ER.

Thigmotropism

The reaction or growth of a plant part to the stimulus of touch is called thigmotropism or heptotropism. Thigmotropic curvature is apparent in the stems of climbers and in tendrils. The entire stem of some vines grows in coils after coils around the supporting object, as is seen in many members of Convolvulaceae. When the stems come in contact with the support, they respond positively by bending towards the object. But some plants such as grapes, peas, *Vitis*, cucurbits etc., have tendrils, which coil around an object to get support and to climb.

Mechanism

The thigmotropic response in tendrils and stems is also due to unequal growth of the organ. Although no direct involvement of auxins has been demonstrated, the hormone might be playing some role as it affects, tendril coiling. The movement of salts and electrolytes is probably involved in thigmotropic responses. In pea tendrils, rapid changes in ATP and inorganic phosphate takes place upon touch stimulus. Perhaps, rapid changes in the permeability of membrane takes place, which besides facilitating the movement of these metabolites, also accelerate the water uptake and cell growth.

Hydrotropism

Response of plant organs to water content is hydrotropism. It can be seen in roots of some species, which show a directional growth towards the region of high water content in the soil. However stems, roots of seedlings of many species do not show any response to the soil water content. But positive hydrotropism has been observed in the growth of pollen tubes.

Chemotropism

Movement by part of a plant in response to a chemical stimulus is called chemotropism. The response by the plant is termed 'positive' if the growth is towards the stimulus or 'negative' if growth is away from the stimulus. Fertilization of flowers by pollen is achieved because the ovary releases chemicals that produce a positive chemotropic response from the developing pollen tube.

HORMONAL CONTROL OF SEEDLING GROWTH

Development is quantitative changes that involve increase in tissue and differentiation of these tissues into various organs of the plant body. The formation of root, stem, leaves etc., is known as morphogenesis. The plant development from seedling to adult plants up to seed production is a complex process which involves various phases like: (i) cell divisions, (ii) cell elongation, and (iii) cell differentiation etc. The different phytohormones regulate or influence a range of various morphogenic, cellular and physiological processes.

Hormonal Control

Hormones are naturally occurring organic substances that, at low concentrations, influence physiological processes such as growth, differentiation and development. Several growth regulators are collectively called the plant hormones or phytohormones. This group includes auxin, cytokinin, gibberellin, abscisic acid (ABA) and ethylene. Many plant hormones are transported from one place in the seedling to another, thus coordinating growth throughout the plant, while others act in the tissue in which they are produced. The hormone must first, bind to a protein receptor, either on the cell surface or inside the cell. This activates a signal transduction pathway which amplifies the signal and to changes in enzyme activities, gene expression and other physiological responses. Plant scientists investigated these phenomena by analyzing the levels and biochemical forms of the hormones present in relation to differences in development. Recently the use of mutants and bioengineered plants in which growth or development is abnormal has enabled us to start understanding how hormones work. This research has been coupled with the isolation of the genes and proteins that are needed for the normal functioning of the hormone.

Biosynthesis

Auxin (indoleacetic acid, IAA) is synthesized from indoleglycerophosphate, the precursor to the amino acid tryptophan, and, in some plants, from tryptophan itself. Gibberellin, is synthesized via the isoprenoid pathway. Cytokinins are synthesized by the attachment of an isopentenyl side chain to adenosine phosphate. Abscisic acid is synthesized via carotenoid molecules. Ethylene is derived from methionine via the nonprotein amino acid ACC (1-amino-cyclopropane-1-carboxylic acid).

Transport

Most hormones simply travel along with the contents of the xylem or phloem by both diffusion and bulk transport. Auxin is special as it is transported primarily in the cells of the vascular cambium initials. Auxin enters the cell from the cell wall above as an un-ionized molecule that can cross the cell membrane. At the neutral pH inside the cell it becomes ionized, preventing its outward diffusion through the cell membrane. Outward transport is on special carrier proteins located only at the base of the cell, so movement is downward (In roots, the situation is reversed).

Signal Transduction

A hormone is a molecule, which must bind to a receptor protein to have its effect. The auxin binding protein is located in the endoplsmic reticulum, from which it also migrates to the cell membrane. After detection, the signal from the hormone molecule has to be translated into action. There are various steps in this process. Usually, the hormone triggers the phosphorylation of an activator protein, which then binds to the regulatory region of a gene, thus turning on gene transcription. This gene may produce the final product, or may itself produce a gene regulator. Prior to the phosphorylation of the regulatory protein may include an interaction with a membrane G-protein that in turn releases other factors and the opening of calcium channels in the membrane permitting an increase in the cytoplasmic level of calcium.

Auxins in Cell Elongation

Indole acetic acid (IAA) is the most widely distributed, naturally occurring auxin in vascular plants. It regulates various processes such as embryo development, induction of cell division, coleoptile and stem elongation, apical dominance, induction of rooting, vascular tissue differentiation at early stage of seedling development, tropic movements etc. Cell elongation is a vital part of growth. Auxin causes cell elongation within ten minutes by making the cell walls more extensible. This occurs through a series of steps. Auxin stimulates the pumping of hydrogen ions out of the protoplast via proton pumps driven by ATP, so acidifying the wall; this activates an enzyme called expansin, which is activated by acid conditions (about pH 4.5). Expansin breaks the hydrogen bonds between the cellulose microfibrils of the wall and the other sugar chain molecules that cross-link the microfibrils; the cell walls are made more extensible; and the cell then elongates because of the turgor pressure inside the cell. Auxin commonly inhibits roots elongation, but high auxin concentrations will promote initiation of secondary, or branch roots.

Role of Gibberellin in α-Amylase Production and Stem Elongation

Two main types of GAs are recognized on the basis of their structure. They are 20 carbon atoms, the C_{20}-GAs and the C_{19}-GAs. Biologically active GAs in plants are C_{19} compounds. The action of gibberellins in the cereal grain germination is the best understood hormonal mechanism in plants. In barley, first gibberellin moves from the embryo to the aleurone layer where the synthesis of α-amylase and protease enzymes takes place. The protease converts an inactive β-amylase to the active form. α- and β-amylase together digest starch to glucose, which is mobilized to the growing embryo for its metabolic activity.

It can be proved by *in vitro* studies. Barley grains may be transected to produce two half-seeds, with one half-seed containing the embryo and one without (Fig. 1.4). The embryo containing half-seed will secrete α-amylase, digest the starchy endosperm, and germinate. The embryoless half-seed neither can germinate nor produce elevated levels of α-amylase. But embryoless half-seed is treated with gibberellic acid will produce high level of α-amylase. Gibberellin is needed for cell and stems elongation.

Cytokinins

The most widespread naturally occurring cytokinin in higher plants is *Zeatin*. Zeatin and other naturally occurring cytokinins are commonly found with a ribose (the riboside) or ribose-phosphate (the ribotide) at the 9th-position. In addition to stimulating cell division, cytokinins also influence shoot and root differentiation in tissue culture.

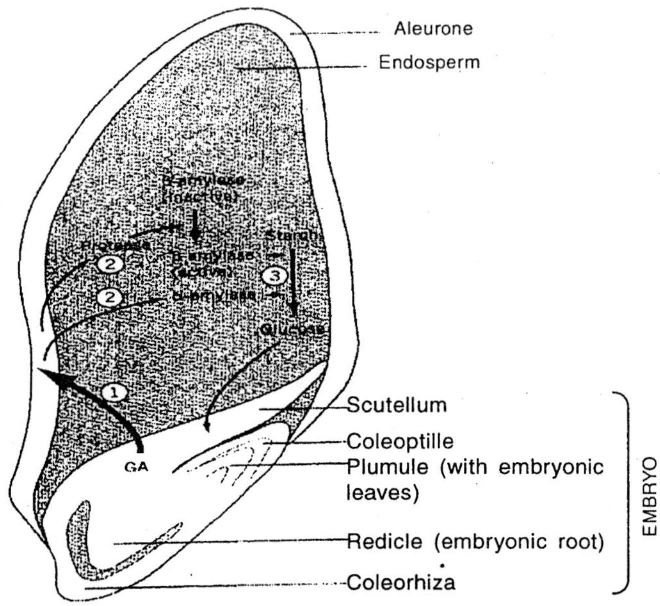

Figure 1.4: A schematic illustrating gibberellin-induced release of enzymes and carbohydrate mobilization during germination of barley (*Hordeum vulgare*) seed.

Abscisic Acid (ABA) and Stomatal Closure

Abscisic acid is a single compound. ABA promotes stomatal closure by causing the exit of potassium ions from the guard cells. K^+ is the main solute causing turgor in the guard cells and opening the stoma. ABA binds to cell-surface receptor on the surface of the guard cells. This causes a calcium ion influx and an increase in the level of inositol triphosphate. The Ca^{++} brings about a membrane depolarization, triggering the outward K^+ ion channels to open.

Ethylene and Seedling Stem Growth

Ethylene is synthesized primarily in response to stress and produced in large amounts by tissues undergoing senescence. Exposure of *Arabidopsis* seedlings to ethylene usually causes stunted growth. Ethylene's receptor interacts with a protein that blocks an ion channel. In the presence of ethylene, the receptor causes the protein to unblock the channel. The entry of (unknown) ions then activates other second messengers. Activation results in the synthesis of a transcription factor, triggering the synthesis of specific enzymes that can cause stunted growth.

GENE EXPRESSION IN SEED GERMINATION

Seed germination begins with water absorption (imbibition). Then the seed food reserves are mobilized. Next the embryo grows by utilizing the mobilized food materials. Mobilization of food reserves varies depending on the nature of the food reserves of the seed. The process has been studied in detail in barley.

Cereal seeds contain a large reserve of starch that is broken down to sugars that supply the developing embryo. The process has been studied in barley. A barley seed has an embryo, an endosperm containing storage reserves, and a layer surrounding the endosperm called the aleurone (Fig. 1.5). If a dormant seed imbibes water, simultaneously the growing embryo produces gibberellic acid. The newly produced gibberellic acid diffuses to the aleurone layer, resulting in the synthesis of new proteins, including starch degrading enzymes such as α-amylase, which break down the starch of the endosperm. Gibberellic acid greatly enhances the rate of transcription of the α-amylase gene. Gibberellic acid response elements (GARE) are present at the transcription site of the gene. Later they aggregate and form gibberellic acid response complex (GARC).

Seed + water → Imbibed seeds → embryo produces gibberellin → gibberellin diffuses to alleurone layer → Gibberellin responsive genes activated → α-amylase produced → which diffuses to endosperm → starch reserves mobilized → which diffuse to embryo.

The gibberellic acid response complex (GARC) is activated by the binding of transcription factors. Proteins are produced in response to gibberellic acid that binds to DNA. Of these GAMYB, is one of the transcription

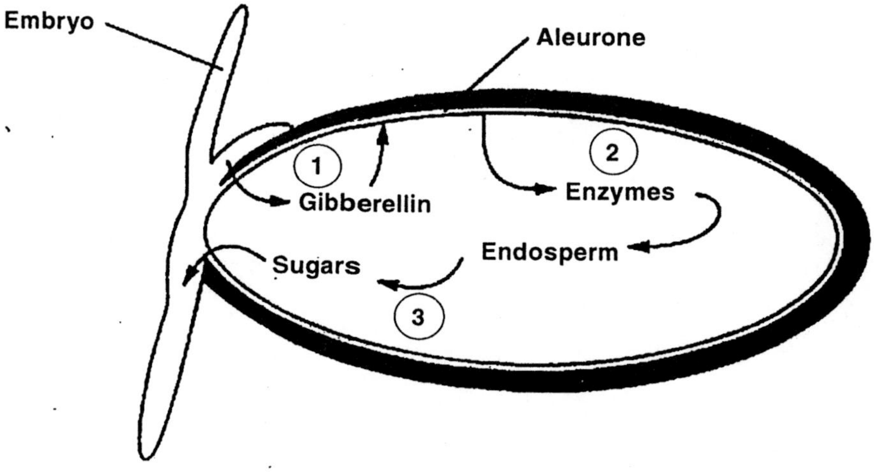

Figure 1.5: Parthway in mobilization of storage reserves in barley grain.

factors (MYBs) known to be involved in development and is upregulated by gibberellic acid well before α-amylase gene expression starts.

During the post germination stages, the shoot can photosynthesize and supply the nutrients required for growth. In grass, production of a fully functional photosynthetic apparatus occurs in three phases. In the first phase, the coleoptile and primary leaf contain etioplasts (precursors of chloroplasts), but that cannot photosynthesize. In the second phase, plastids go on dividing and increase in abundance. Genes are activated which generate the photosynthetic apparatus and enzymes of the Calvin cycle. This involves activity of genes of both the plastid genome (plastome) and nuclear genome and the efficiency of the photosynthetic system increases as more components are added. In the third phase, the activity of the photosynthetic apparatus is maintained, and is repaired, if the damage occurred.

CHAPTER II

CELL, CYTOSKELETON AND ERGASTIC SUBSTANCES

INTRODUCTION

The cells are the basic units with which the organisms are constructed. Robert Hooke (1665) gave the term 'Cellula' for the first time to the small cavities surrounded by walls in cork. Later, he observed similar cells in other plant tissues and they also contained 'Juice' (Matzke, 1943).

The protoplasm — the substance with in the cell was discovered later. Hanstein (1880) used the term 'protoplast' to indicate the unit of protoplasm found in a single cell. The term 'cell' in plants includes the protoplast together with the wall. Until very long, the cell wall was considered as a non-living excretion of the living cell matter. But, recently, more and more evidence has been found that organic unity exists between the protoplast and the wall, and both of them together form a single biological unit.

Robert Brown (1831) discovered the nucleus. Kolliker (1862) introduced the term cytoplasm. Lot of research has been done during 19th and 20th centuries on cytoplasm and nucleus and it developed into a branch of science known as cytology. Prokaryotic and eukaryotic cells are distinguished on the basis of internal organization. The eukaryotic cells are present in all plants and animals except bacteria and blue-green algae and are characterized by a membrane bound nucleus and other organelles. Prokaryotic cells lack those things.

More and more knowledge has been accumulated on the structure and biology of the cell after the discovery of the electron microscope. The protoplast consists of membranous and non-membranous components. When observed under an electron microscope, the membranes are seen as two dark lines each about 2.5 nm thick, separated by a lightly stained line of 3.5 mm thick. This type of membrane consists of a bimolecular lipid layer covered on each side with a layer of protein

The main components of plant cell are the cell wall, cytoplasm and nucleus. The cytoplasm includes the endoplasmic reticulum, Golgi apparatus, mitochondria, plastids, microbodies, ribosomes, spherosomes, microtubules, vacuoles and ergastic substances. The cell generally contains a single nucleus, but in some cells like the sieve elements of the phloem, the nucleus is absent. There are also cells like some laticifers with numerous nuclei in them. Such a structure is called a coenocyte. The coenocyte aroused lot of interest in phylogenetic and ontogenetic studies. Two theories are there which deal with the relation of the entire organism and the single cell. According to 'cell theory', the organism consists of an enormous number of cells each of which plays a role in determining the nature of the organism. The second theory namely 'organismal theory' contradicts the cell theory. The organismal theory gives less importance to the individual cells and gives more stress on the unity of the protoplasmic mass of the entire organism. According to this theory, the organism as a unit, determines the nature of the cells. Both these theories are important.

THE CYTOSKELETON

The cytoskeleton is involved in the maintenance of cell shape. It plays an important role in cell mobility and the colloidal changes that the cytoplasm may undergo. The microtubules and a variety of microfilaments and microtrabeculae present in the cytoplsmic matrix form a kind of dynamic and spongy cytoskeleton. It provides a frame work for soluble proteins, enzymes and ribosomes.

The name cytoskeleton was coined many years ago, but it was abandoned because many structures observed in cytoplasm with light microscope were considered to be fixation artifacts. The existence of an organized fibrous array in the protoplasm was postulated by Koltzoff in 1928. He opined that "each cell is a system of liquid components and rigid skeletons, which generate the shape, and even though we rarely see the skeletal fibrils in living and fixed cells, that only means that these fibrils are very thin or that they are not distinguished by their refractive index from the surrounding colloidal solution". Thus, he conceived of a cytoskeleton which determines both the shape of the cell and the changes in its form. Electron microscope have confirned this assumption and revealed that the cytoplasm of most eukaryotic cells contain a cytoskeletal basic fabric formed of microtubules, microfilaments and intermediate filaments. These microtubules, microfilaments and intermediate filaments are related to the various forms of cell motility. The main proteins that are present in the cytoskeleton are tubulin, actin, myosin, tropomyosin etc.

Microtubules: These are the structures commonly present in the cytoplasm of eukaryotic cells. These are characterized by their tubular appearance and their uniform properties in the different cell types. De Robertis and Franchi (1953) first observed these tubular structures in the axoplasm extruded from myelinated fibres. These so called microtubules appear as elongated, unbranched, cylindrical elements of 20–30 nm in diameter and indefinite length. Microtubules were observed in a variety of animal cells. Cytoplasmic microtubules are uniform in size and are straight. In a cross section, it shows an annular configuration with a dense wall of about 6 nm thick and a light center. Each microtuble is surrounded by a zone of low electrón density. All the microtubules studied shown almost the same morphological characteristics but differ in other properties. For example, the microtubules of cilia and flagella are more resistant to various treatments. Cytoplasmic microtubules nomally disppear if stored at 0°C or after treatment with colchicine.

Tubulin — The Main Protein of Microtubules

Microtubules are composed of protein sub units which are similar though they are found in different cell types. The term tubulin is the principal protein of cilia and flagella and is also the protein of cytoplasmic microtubules. Tubulin is a dimer of 110,000 to 120,000 daltons. Two different monomers — Tubulin A and B are present in flagella. Mostly, tubulin is a heterodimer having two monomers of different kind. The amino acid composition of tubulin from different sources shows little variation. Some enzymatic activities like proteinkinase activity have been reported for tubulin.

Microtubules can be assembled *in vitro*. For this, presence of GTP (Guanosine triphosphate) and high concentrations of tubulin, as well as some associated proteins are necessary. The polymerization is both temperature and pressure sensitive. When microtubules are disassembled, a mixture in the form of rings is produced. These rings are composed of tubulin dimers and associated proteins. The role of these rings in microtubule assembly is questionable because they are not observed *in vivo* and appear only *in vitro*. The assembly-disassembly process is a dynamic equilibrium (Fig. 2.1).

The assembly of tubulin in the formation of microtubules is a specifically oriented and programmed process. Centrioles, basal bodies and centromeres are sites of orientation for this assembly. Calcium and calmodulin may be regulating factors in the *in vivo* polymerization of tubulin. The level of polymerized tubulin is high at interphase and metaphase (i.e., spindle microtubules) and lower at prophase and anaphase. Assembly involves the hydrolysis of GTP to GDP. The polarity of the assembly can be demonstrated, under the Electron microscope, by the addition of tubulin.

Cell transformation (i.e., cancer) produces a disorganization of microtubules. Several functions have been attributed to microtubules. (1) Cell motility in primitive forms. (2) Microtubules play a mechanical function. The shape of the cell and cell processes is dependent on microtubules. This function is apparent during cell differentiation of certain placodes, in the

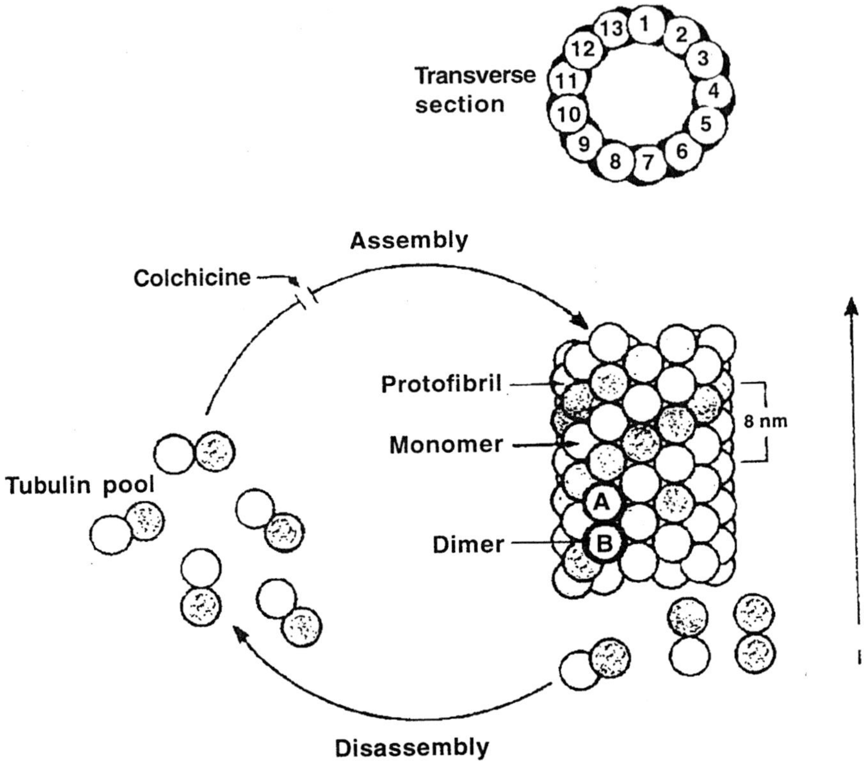

Figure 2.1: Diagram of a microtubule and the process of assembly and disassembly of tubulin. *Top*, transverse section of a microtubule showing 13 protofibrils. *Bottom*, longitudinal view showing the helicoidal protofibrils made of tubulin dimers and the polarization of the microtubule. Observe that the microtubule is being disassembled at the bottom while being simultaneously assembled at the top. Colchicine, by blocking the assembly process, produces depolymerizatiön of the microtubules.

nerve cells and in spermatogenesis. (3) The polarity and directional gliding of cultured cells depend on microtubules. (4) Microtubules are also associated with transport of molecules, granules, and vesicles with in the cell. (5) They play a role in the contraction of the spindle and movement of chromosomes and centrioles as well as in ciliary and flagellar motion. (6) A possible role in sensory transuduction also has been postulated.

Microfilaments: Thin microfilaments ranging between 5–7 nm in width represent the active or motile part of the cytoskeleton and appear to play the major role in cyclosis and amoeboid motion. With high voltage Electron microscopy a three-dimensional view of microfilaments has been obtained (i.e., an image of microtrabecular lattice). These microfilaments are sensitive to cytochalasin-B, an alkaloid that also impairs many cell activities such as the beat of heart cells, cell migration, cytokinesis, endocytosis and exocytosis. Applied to the alga *Nitella*, cytochalasin-B produces a rapid halt in cyclosis. It is generally assumed that the cytochalasin-B sensitive microfilaments are the contractile machinery of non-muscle cells.

The contractile proteins such as actin and myosin, as well as tropomyosin, troponins-e, a-actinin and other proteins found in muscle, are present in microfilaments. These proteins can be localized with specific antibodies. Actin filaments are made of globular actin molecules (G-actin). In the presence of myosin they become 'decorated' as can be seen with the electron microscope.

Actin is one of the most abundant proteins in many cells. The globular (G-actin) fibrillar transition is the basis of the classical sal-gel transition in the cytoplasm of moving cells. Myosin is found in amoebae, blood platelets and slime molds but in a much lower concentration than actin. It contains a Ca^{++} activated ATPase as is found in muscle myosin. Identification of myosin filaments is more difficult.

Intermediate Filaments

In addition to the microtubules and the microfilaments, in the cytoskeleton of many cells, there is a group of 10 nm filaments called intermediate, because they range in thickness between the microtubules and the microfilaments. These are highly heterogeneous in their biochemical properties. By their morphology and localization, these filaments can be grouped into four main types namely neuron filaments, glial filaments, keratin filaments and a more heterogeneous type, which is composed of several proteins.

In spite of their heterogeneity, intermediate filaments show a similar structural organization. When seen with X-ray diffraction, all members of this family contain considerable amount of d helix. When examined under the EM after shadow casting, show a periodic structure with a length of 21 nm. Intermediate filaments are insoluble in physiological solutions but dissolve at low and high pH. After dissolution they can be reconstituted again to form intermediate filaments.

Intermediate filaments form networks that interconnect the nucleus with the cell surface. In fibroblasts intermediate filaments attach to the nuclear envelop and form a case around the nucleus; at the plasma membrane they appear associated with microfilaments at the adhesion plaques. The intermediate filaments that form the nuclear case are continuous with other intermediate filaments that radiate through the cytoplasm. Intermediate filaments are formed by protein subunits of a multigene class that are expressed differentially in various cell types. However, the different intermediate filament proteins are organized in a similar way forming tetrameric complexes.

(1) Keratin filaments, also known as tonofilaments, are present in epithelial cells and contain the α-keratins. (2) Neuro filaments are present in

axons, dendrites and perikarya of neurons. (3) Glial filaments, containing a very acidic protein, are present in astrocytes. (4) Heterogeneous filaments have a similar morphology and localization but contain a variety of proteins. This group consists of the desmin filaments, which, in muscle, constitutes a cytoskeletion that holds the myofibrils within the muscle cell. This cytoskeleton also contains vimentin and synemin. Vimentin filaments are wavy and in many cells tend to form perinuclear caps.

Intermediate filaments are resistant to clochicine and cytochalasin-B and are sensitive to proteolysis. They are involved in mechanical functions contributing to the determination of the cell shape and to the integration of cellular compartments. During mitosis of some epithelial cells, intermediate filaments undergo disintegration and restoration

ERGASTIC SUBSTANCES

Plants synthesize and accumulate a variety of compounds in them. Some of these compounds have high nutritional value like starches, proteins and fats but some were considered to be 'waste products'. During recent times, there is a tendency to avoid/ dismissing compounds as mere waste products, with no value. Harborne (1982) opined that many of these ergastic substances which were once believed as 'waste products' are not really 'waste', but they are useful to the plants in many ways. These ergastic substances makes the plants, either less nutritious or bitter and unpalatable and deter the forage animals from eating them. A few of these ergastic substances are crystalline in nature and effectively inhibit the insects to chew the plant parts or lay eggs in them.

The chemical composition of the ergastic substances seen in the histological / anatomical preparations was not known quite a long time and similarly the advantages they confer to the plants. But, with the advent of several new techniques recently, it is possible to analyse those small amounts microscopically and know their chemical nature and properties. The ergastic substances can be broadly classified as:

Ergastic Substances

- Food products
  - Non-Nitrogenous: Starch, Inulin, Hemicellulose, Cellulose, Sugars
  - Nitrogenous: Proteins, Amino-compounds
  - Fats and fatty oils
- Secretory products
  - Enzymes, Pigments, Nectar
- Waste products
  - Non-Nitrogenous: Tannins, Mineral Crystals, Latex, Essential Oils, Gums
  - Nitrogenous alkaloids

Food Products

These are the substances manufactured by the protoplasm of the cells by using simple inorganic substances like CO_2, H_2O and stored in the cells as food materials. The food is partly used to make up new protoplasm and is partly broken down to provide the energy required for performing the normal activities and the remainder is stored as reserve food material in the protoplasm of the cells. The reserve food materials can be classified as: (a) the carbohydrates, (b) nitrogenous products, and (c) fats and fatty oils.

Carbohydrates

These are the products more or less directly derived from carbon dioxide and water during 'photosynthesis'. Some of the carbohydrates are insoluble while some are soluble in water. The most important insoluble product is the starch and the soluble products are inulin and sugar.

Starch

This is a carbohydrate which is composed of long chain of molecules. It appears in the form of grains which are formed in the chloroplasts. Later, the starch is broken down into sugars which then moves into the storage tissues where it is resynthesized in the amyloplasts. Starch grains are abundantly found in tuberous roots, underground stems, cortex of stems, endodermis, grains of cereals, fruits like banana etc. (Fig. 2.2). Starch grains vary in their shape and size and is used as an identification feature of a taxa i.e., oval starch grains in potato, flat grains in wheat, polygonal in maize, spherical in pulses, dumbell or rod shaped in the latex of Euphorbias. Starch grains of rice are the smallest and those of *Canna* are the largest. The variation in size ranges from 5–100 µ.

Starch grains show layering around a point termed 'Hilum'. It is situated at the center in *Triticum* or may be eccentric as in Potato. The number of layers in the starch grains of cereals corresponds to the number of days during which it grows. The commonest type of starch grain is concentric. If the starch grain possess a single hilum it is known as simple. Sometimes, 2, 3 or many grains are arranged in groups and had many hila, known as compound grains, e.g., potato, rice, oats etc. Starch grains turn to blue or black when aqueous Iodine solution is added and appears shining or luminous under a microscope except for dark cross.

Inulin: This is also a carbohydrate of polysaccharide type. As it is a soluble carbohydrate, it is found in the cell sap of roots of many Asteraceae. It is common in tuberous roots of *Dahlia* and *Helianthus*. Inulin can be obtained in the form of precipitate by keeping the roots of *Dahlia* in alcohol for 6–7 days. It appears in the form of spherical, star shaped or wheel shaped crystals.

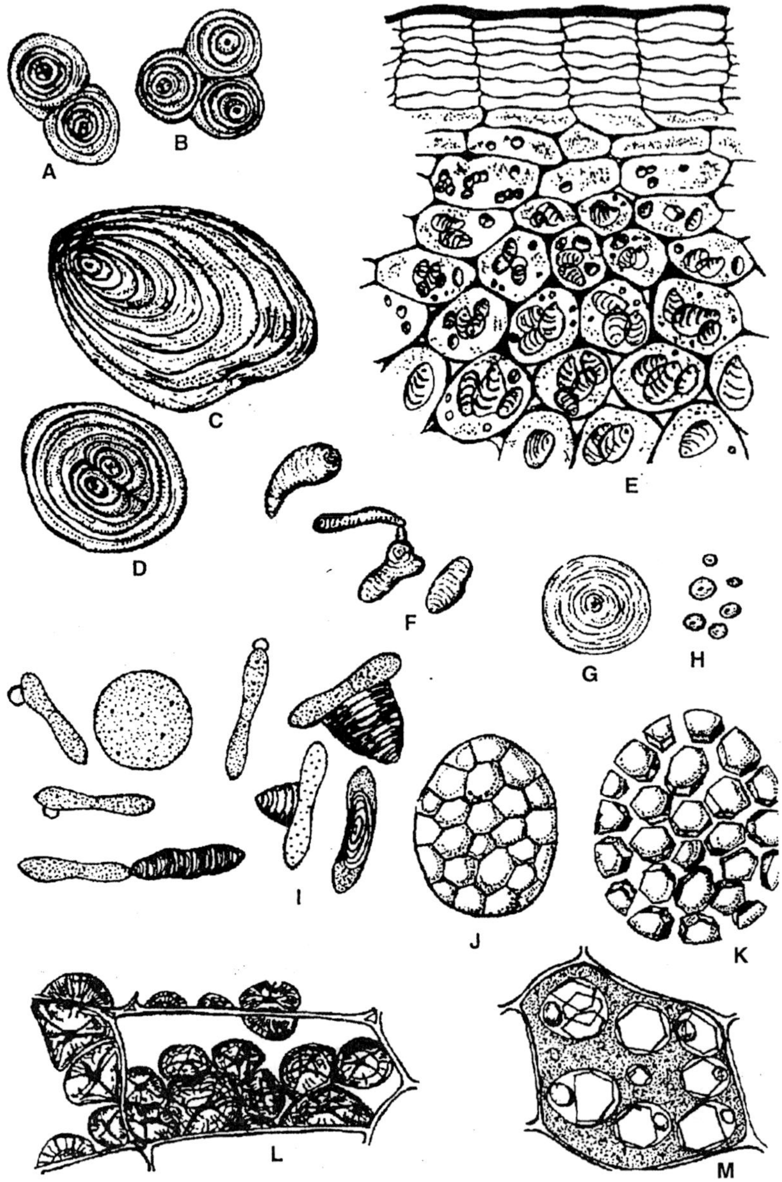

Figure 2.2: (A–D) Potato starch grains; (A, B) Compound starch grains; (C) Simple starch grain; (D) Half-compound starch grain; (E) Cross-section of the outer portion of a potato tuber; (F) Banana starch grain; (G, H) Starch grains of *Triticum durum*; (I) Stages in development of starch grains in chloroplasts of *Phaius maculata*; (J) Compound starch grain of *Avena*; (K) As in J but disintegrating; (L) Sphaerocrystals of inulin in cells of *Dahlia* tuber, precipitated when alcohol was added; (M) Aleurone grains in an endosperm cell of *Ricinus communis* from a section of material embedded in dilute glycerine.

Hemicellulose: The food is stored in the form of hemicellulose in some seeds. This type of storage is rather rare when compared to starch or sugar. Hemicellulose (reserve cellulose) is found in the seeds of certain palms.

Cellulose: This is also a carbohydrate having a similar formula like that of starch i.e., $C_5H_{10}O_5$. However, the arrangement of the atoms in the molecule is quite different in starch and cellulose hence they have different properties. In plants the cellulose is made from sugars and serves as a building material in the formation of a cell wall.

Sugars: The common sugars that are found abundantly are glucose, fructose and sucrose. Glucose and fructose had a similar formula i.e., $C_6H_{12}O_6$. They are composed of same kinds of atoms, in the same proportion, but the arrangement in the molecule is different. These are monosaccharides and are the simplest of the natural sugars. Sucrose is a disaccharide and has twice as many carbon atoms and the formula is $C_{12}H_{22}O_{11}$. Sucrose is known as the ordinary sugar and is obtained from sugarcane or sugar beets. The sugars are the simplest carbohydrates and are quite soluble in water. Glucose and fructose have relatively simple molecules. They are good molecules for the building up of other substances or for furnishing energy. In sugarcane or sugar beet sucrose is found as reserve food.

Nitrogenous Products

Proteins and amino compounds are the two important nitrogenous compounds:

Proteins: The proteins are the most important compounds found in plants. They constitute the active matter of protoplasm as many chemical phenomena of the life processes are associated with them. The proteins are highly complex organic nitrogenous substances consisting of carbon, hydrogen, oxygen and nitrogen. In some more complex proteins, sulphur and phosphorus are also present. For example, protein Zein from Indian corn has the formula $C_{736}H_{116}N_{184}O_{208}S_3$, Gliadin from wheat $C_{685}H_{1068}N_{196}O_{211}S_5$. In plants, the proteins are found frequently as solid granules. Proteins are formed due to rearrangement of atoms of carbohydrates with the addition of nitrogen and sulphur commonly and sometimes phosphorus also. A protein molecule is made up of 100s or 1000s of amino acid molecules joined together by peptide links into one or more chains which are variously folded. 20 different kinds of amino acids are commonly found in proteins. Most of these amino acids generally occur in any protein molecule. They are arranged in a chain sequence which is exactly the same in all molecules of a given kind of protein. The different arrangements of amino acid molecules is practically infinite. Every species have protein molecules peculiar to itself. The proportion of different amino acids in different proteins varies. Some proteins lack the amino acids found in other proteins. Animal proteins are better as human food than plant proteins because the amino acids of animal proteins are more like that of human proteins. Some plant proteins lack some of the amino acids which are essential for the building of human proteins.

A common form of protein found in the endosperm of castor seed is known as aleurone grains. Each aleurone grain is a solid, ovate or rounded body which usually encloses a crystal like body in it known as crystalloid and a globule like body known as globoid. The crystalloid is proteinaceous in nature and occupies the major portion of aleurone grain, where as the globoid is double phosphate of calcium and magnesium which occupies the narrower part of the grain. The aleurone grains vary in their shape and size in different plants. When they are found in oily seed of castor, they are larger in size, while when found with starch, they are very minute in size.

Amino compounds: These are simple nitrogenous food materials. They are found in the form of amino acids and amines which occur in the cell sap. These are found abundantly in the growing apices of the plants and less frequently in storage tissues.

Fats and Fatty Oils

The fats and fatty oils in plants are composed of glycerin and organic acids. They occur in the form of minute globules in the protoplasm. Special types of fats and fatty oils are found in the seeds and fruits of flowering plants. Fats and fatty oils are composed of carbon, hydrogen and oxygen. They are characterized by a small proportion of oxygen as can be seen from the formula of common fats, e.g., palmitin $C_{15}H_{98}O_6$, olein $C_{57}H_{104}O_6$ etc. Because of the very small percentage of oxygen contained in fats, the oxidation of fats produces large quantities of energy. Fats are insoluble in water but soluble in ether, chloroform and petroleum. Fats may be solid or liquid (fatty oils) according to temperature. Fats and oils may be produced by elaioplasts or spherosomes. Their distribution in plants is restricted than that of starch. Lipids occur as spherical droplets and can be easily detected. Lipids readily absorb stains like sudan III and sudan IV. Rarely lipids also occur in crystalline state, e.g., needle shaped fat crystals in the endosperm of the palm — *Elaeis*.

SECRETORY PRODUCTS

Besides food materials, several other products are also secreted by the protoplasm which are not useful as nutritional products, but they may help or accelerate the various reactions in the cells. They are:

Enzymes: These are soluble nitrogenous substances secreted by the protoplasm. They are digestive in function and convert the insoluble substances into soluble and complex compounds into simpler ones, e.g., the enzyme diastase converts the starch into sugar. Because of the action of this enzyme an insoluble substance is converted into soluble one. The enzyme lipase breaks fats into glycerine and fatty acids. Papain converts proteins into amino acids.

Pigments: The plant pigments are usually found in the plastids and in the vacuole. The green colour of the leaf is due to chlorophyll which is found in the chloroplasts. In the same plastids, carotenoids — the yellow to red pigments are also found, but they are masked by the chlorophyll. The carotenoids become noticeable when there is little or no chlorophyll as is the case in chromoplasts. Chromoplasts are also developed from proplastids Another group of pigments, flavonoids (anthocyanins and flavones or flavonols) are generally present in vacuoles. They are water soluble and give various colours to many flowers and fruits. Anthocyanins give red, pink, lilac and blue colour. This depends on the pH (acid solution-orange red to lilac, base pH 7.0 blue).

Nectar: Nectar is another useful secretion of the protoplasm secreted by special glands called nectaries. If they occur on the flower they are called floral nectaries, and those occurring on vegetative parts are called extrafloral nectaries. Floral nectaries are found on various positions of the flower and extrafloral nectaries occur on stems, leaves, stipules etc. The secretory tissue of the nectary is found on the epidermis or a few subepidemal layers. The nectary is covered by a cuticle. The sugar of the nectaries (both floral and extrafloral) is derived from the phloem. Vascular tissue is found near the nectary. The nectar is excreted either through the cell wall and the ruptured cuticle or sometimes through stomata.

Osmophors: The fragrance of the flowers is generally produced by volatile substances distributed through out the epidermal region of the perianth parts. In some plants, the fragrance originates in special glands known as 'osmophors'. In Asclepiadaceae, Aristolochiaceac, Araceae, Burmanniaceae and Orchidaceae, various floral parts are differentiated as osmophors and they assume the form of flaps, cilia or brushes. The prolongation of the spadix of Araceae and the insect attracting tissues in the flowers of Orchidaceae are osmophors. The osmophors have secretory tissue which is several layers in depth.

Waste Products: These are the excretions of the plant cells. These products are stored in the dead cells. They are formed as a result of metabolic activities of protoplasm and hence are known as metabolic wastes of the plants. There are two types of metabolic wastes. They are: Non-nitrogenous waste products and nitrogenous waste products.

NON-NITROGENOUS WASTE PRODUCTS

These are of various types like tannins, mineral crystals, latex, essential oils, gums, resins and organic acids. They are found in the cytoplasm or cell sap.

Tannins: These are heterogenous group of phenol derivatives. In microscopical sections they appear as granular masses which are coloured as yellow, red or brown. Tannins are found in different parts of the plant like leaves, periderm, vascular tissue, unripe fruits, seed coats etc. These are found in individual cells (Idioblasts) or in special organs called 'tannin sacs' within the cell. The tannins are also found in the vacuole, or in the form of droplets in the cytoplasm. Sometimes, for instance in the cork tissue, they penetrate into the cell walls. Tannins are believed to protect the plants against dehydration, rotting and damage by the animals. The tannins can deter the insects feeding on the plants by interacting with and damaging the proteins of the gut. Tannins, as substances associated with starch metabolism, are useful in the transport of sugars and also acts as antioxidants. Tannins are commercially useful in the tanning industry i.e., tanning of animal skins to obtain leather.

Mineral crystals: Various types of mineral crystals occur in plant cells. They occur in the cell cavity or in the cell walls. The crystals may lie loose in the cells or suspended into cell cavity from the cell walls. The crystals vary in shape and size. The crystals consists of calcium carbonate or calcium oxalate or silica.

Calcium carbonate crystals: These are commonly known as the cystoliths. They are present in the epidemal cells of leaves of many flowering plants like Moraceae. Urticaceae, Acanthaceae, Cucurbitaceae etc. The main body of a cystolith is a cellular extension of the cell wall in which the calcium carbonate is deposited in the form of granules. Sometimes, double cystoliths are also found. These cystoliths project into the protoplasts in large, specialized cells. They vary in shape in different genera and families.

Calcium Oxalate crystals: The leaves and other organs of many plants contain conspicuous crystals of calcium oxalate. The forms of these crystals are very diverse.

Raphides: These are long, needle like crystals lying parallel to each other in a bundle. which sometimes found in a special sac like cell. When these sacs are mechanically injured, these raphides are released with jerk through a hole. The raphides are commonly found in *Alocasia, Colacasia, Pistia* etc. These raphides are very irritating and afford some protection to the plants from the forage animals. The raphides can be destroyed by boiling. Hence the food plants containing them are not irritating when cooked.

Druses, Rosette crystals or cluster crystals: This is one of the commonest crystal and had the rosette shape hence the name rosette crystal. These are quite common in *Eucalyptus, Nerium* etc.

Prismatic crystals: These are single calcium oxalate crystals found in various plants. They may be square prismatic, rhomboidal or pyramid like in shape.

Sand crystal: These are commonly found in masses of microspheroidal crystals packed into a cell. These are found in the leaves and roots of *Atropa belladona*. Silica is deposited mostly in the cell walls and sometimes in the lumen of the cell. Members of the family Poaceae are the best example of a plant group with silica both in the cell walls and the lumen. The term silica body is used for the silica deposits of Cyperaceae and Gramineae.

Stegmata: These are restricted to epidermis or cells adjascent to vascular tissues only. Stegmata are valuable diagnostic character of the members of the group Palmae.

Latex: This is the milky or watery juice found in the long branching tubes known as latex tubes. The neighbouring latex tubes get connected and form a net work. The latex oozes out when these tubes are cut or injured. Rubber, opium, chewing gum are the valuable substances derived from the coagulated latex. The latex usually contains the starch grains, proteins, oils, tannins, gums, resins, alkaloids, salts, enzymes and also some poisonous substances.

Best known latex is the one obtained from rubber yielding plants. The latex of *Hevea brasiliensis* consists of 40–50% of the rubber. Rubber particles, suspended in latex vary in size and shape. When the latex is released from the plant, the particles clump together i.e., the latex coagulates. This property is utilized in the commercial separation of rubber from latex. The latex may be transparent or clear (*Morus, Nerium*) or milky (*Asclepias, Calotropis, Euphorbia, Ficus* etc.), yellow-brown in *Cannabis*, yellow or orange in some members of Papaveraceae.

Essential oils or Volatile oils: These are volatile, very odoriferous and frequently occur in oil glands as in *Eucalyptus*. The orange peel contains large, oval glands which are formed by the breaking down of the cells containing the oil. The disintegration of the cells brings the oil into large cavity of the gland. Oil glands are also found in the petals of rose, jasmines and other flowers.

Gums: The gums are exuded from the stems of many plants. *Acacia* species like *A. senegal*, *A. modesta* and *A. arabica* produce the best gum. The gums are soluble in water and insoluble in alcohol. The gums swell up in water and form viscous mass. These are related to peptic compounds. They appear in plants mainly as a result of physiological or pathological disturbances which induce break down of cell walls and contents.

Resins: These are frequently found in the resin ducts of conifers. Resin ducts are formed either by the separation or disintegration of cells. The turpentine is obtained by cutting of resin ducts of pine trees. Various types of varnish and other resins are obtained from other conifers. Resins are insoluble in water, but soluble in alcohol. The well known mounting fluid canada balsam is a resinous product of *Abies balsamea.*

Resin Ducts: These are long passages surrounded by a glandular cells. These occur in stems and also in other parts of plants. In *Pinus,* resin ducts develop in the axial and ray system. Resin ducts develop as schizogenous intercellular spaces by the separtion of parenchyma cells from each other. Some parenchyma cells form the lining or epithelium over the ducts and excrete resin. In *Pinus*, the epithelial cells are thin walled, active for several years and produce abundant resin. In *Abies* and *Tsuga,* the epithelial cells have thick, lignified walls and most of them die during the year of origin. They produce little resin. Eventually, the resin duct may become closed by the enlarging epithelial cells. There is distinction between normal resin ducts and those that are traumatic (trauma = wound) i.e., arise in response to injury. Normal ducts arise singly where as the trauamatic ducts are cyst like and occur in tangential series.

Organic Acids: These are found in several vegetables and fruit juices and re-combined with peculiar bases and alkaloids. Several organic acids are found in various plants viz., citric acid in citrus fruits, tartaric acid in grapes and tamarinds; malic acid in the fruits of apple, gallic acid in Mango seeds, Oxalic acid in *Rumex, Oxalis, Nepenthes* etc.

NITROGENOUS WASTE PRODUCTS

Alkaloids: These are a group of nitrogen contining basic organic compounds present in plants of a few families of dicots like Solanaceae, Papaveraceae etc. They are thought to be the end products of nitrogen metabolism. These are important because of their poisonous and medicinal properties, e.g., Atropine found in *Atropa belladonna* of Solanaceac, cocaine, morphine, nicotine, quinine, strychnine etc.

CHAPTER III

CELL WALL

Plant cells possess cell wall and this feature distinguishes them from animal cells. Cell wall is the non-living component of the cell and is synthesized by protoplasm. The synthesis of cell wall is controlled by Golgi bodies. The cell wall is protective in function.

The cell wall was discovered in 17th century even before the recognition of the protoplast. Since then several studies have been made by using various methods — chemical, physical and morphological. These investigations made use of the advances made in organic chemistry, X-rays, the use of the Light, Polarizing and Electron Microscopes and helped in the better understanding of the origin and structure of the cell wall.

The cell wall is in contact with the protoplast and grows outside of it. Only in the pollen grains the outermost wall — exine develops from the contribution of both i.e., by the protoplast within and the tapetum which surrounds the pollen grains.

The cell walls vary in thickness basing on their age and type of cells. Generally, the younger cells have thinner walls than the older ones. In mature thick walled cells, a concentric layering is usually distinct in the cell wall. These layers differ from one another in their physical and chemical nature. Three principal layers are found in the cell wall (i) The middle lamella, (ii) The primary wall, and (iii) The secondary wall.

The middle lamella: It is the cementing layer which binds the two adjacent cells in a tissue. It is therefore, present between two primary walls of neighbouring cells. It is an amorphous substance, inactive in terms of polarized light (isotropic). In supporting tissues, it also fills intercellular spaces. The middle lamella consists of mainly pectic substances and usually impregnated with calcium and magnesium. Pectic substances are a mixture of polygalacturon (D-galacturonic acid) and polysaccharides (neutral sugars). The enzyme pectinase and chemical reagents which solubilize pectins, disintegrate the tissue into individual cells. The procedure is called maceration.

The formation of this layer starts in the telophase stage of cell division and is formed by deposition.

The primary wall: Bordering middle lamella the primary wall is the first layer of individual cell, synthesized by protoplasm. It is 1–3 μm thick. The primary wall consists of hemicellulose (up to 50%), cellulose (up to 25%) and smaller amounts of pectic substances, non cellulosic polysaccharides (xylans, mannans, galactans, arabinogalactans, glucomannans) and forms the matrix of wall in which cellulose microfibrils are embedded. The primary cell wall is elastic and undergoes extension with the growth of the cell wall. In many roots, fleshy stems, fruits and leaves the cells contain only the primary cell wall and the middle lamella. The primary wall is not uniformly thickened but it may be alternatively thickened or thinned. The changes that occur in this wall are reversible. The walls may lose a thickening previously acquired and the chemical substances may be removed or replaced by others. For example, cambial walls show seasonal changes in thickness. The primary walls of the endosperm cells are digested during germination. The primary wall is optically active (anisotropic).

Secondary wall: It is formed next to the primary wall towards protoplasm. It appears usually after the cell caeses to grow. Hence, surface growth is normally absent in it. It is about 5–10 μm thick. It is very strongly anisotropic and is characteristic of highly specialised cells which undergo irreversible changes in their development. Rarely, the active cells with living protoplasts like xylem ray and xylem parenchyma cells also have secondary walls. The secondary wall is laid down over primary wall except over pit membrane. The secondary wall is more massive than primary wall and in most cells it forms the major part.

In majority of tracheids and fibres three layers can be observed in secondary wall the outer layer (S_1), the central layer (S_2) and the inner layer (S_3). Of these layers, the central is usually the thickest. Sometimes new term-teritary wall or terminal layer is applied to the inner layer (S_3). The teritary wall (S_3) also differs from the other layers of the secondary wall in its chemical composition. The S_3 layer is covered by a noncellulosic film often with lumps called warts and known as warty layer. This may be formed by the remnants of a disorganised protoplast, e.g., conifer tracheids and fibres and vessels of many dicotyledons. In some cells, the number of layers may be more than three. Some authors use the term tertiary wall for the innermost layer of the secondary wall. According to Frey-wyssling (1976) the innermost lamella (tertiary lamella) differ with secondary wall in some properties. It is differentiated into two strata — a membranogenic stratum and a warty stratum. The secondary wall is composed of cellulose, hemicellulose and lignin. Because of high cellulosic content, it is strongly anisotropic. Hemicellulose is less in the secondary wall when compared to the primary wall. The secondary wall provides mechanical strength to the plant body and is usually associated with non-living tissues at maturity. It is also characterstically found in cells which undergo irreversible changes in their development.

FORMATION OF THE CELL WALL

During the telophase, of the mitosis, the phragmoplast widens and becomes barrel shaped. At the same time, on the equatorial plane, the cell plate, i.e., the first partition between the new protoplasts begins to form inside the phragmoplast. In the area where cell plate forms, the microtubules of the phragmoplast disappear, but successively regenerated at the circumference of the cell plate. With the enlargement of the cell plate, the microtubules approach the wall of the cell. With Electron microscope it has been shown that the cell plate formation is initiated by the concentration and fusion of large number of vesicles derived by the golgi bodies. ER vesicles also may be involved in this. This microtubules of the phragmoplast seems to be involved in directing the vesicles towards the equatorial region. Individual plasmatic linkages remaining in the gaps of the cell plate form the plasmodesmata.

On both sides of the middle lamella, thin lamellae are laid down by the daughter protoplasts. Formation of these lamellae is the starting point in the development of the new walls of the daughter cells. The walls consists of cellulose microfibrils and non-cellulose matrix. The matrix consists of pectic substances and hemicellulose. This wall matrix is secreted by the golgi vesicles. In some cases ER also contributes in the production of matrix.

Along the line of contact of the new wall and the wall of the mother cell, the new and old middle lamellae are separated by the primary wall of the mother cell. In the primary wall of the mother cell, a cavity which is triangular in cross section develops along the line of contact of the new and old walls. This cavity continues to enlarge till it reaches the middle lamella of the mother cell and so connection is made between the new and old middle lamellae. If the cavity continues to grow and the intercellular substance does not fill it, an intercellular space lined with intercellular substance is formed.

Fine structure of cell wall: Usually cell wall components are, cellulose, hemicellulose, pectin, lignin, suberin, cutin, proteins etc. The proportions of these substances differ from species to species. Of these, the architecture of the cell wall is mainly based on cellulose. The cell wall is composed of macrofibril, 0.5 μm in width, which are visible under the light microscope. The macrofibril consists of a bundle of microfibrils, each of which is about 250Å in diameter. The microfibril is visible only under the Electron microscope. The microfibril in turn consists of a bundle of micelle or elementary fibrils, each about 100Å in diameter. The micelle contains about 100 cellulose chains. Cellulose is a polysaccharide with linear chains of glucose ($C_6H_{10}O_5$). In the cell wall, cellulose forms the framework interpenetrated by the matrix representing the non cellulosic substances (e.g., lignin).

Micelles are composed of approximately 100 individual cellulose chains with cross-sectional area of about 3000Å. The molecules of cellulose, however, consists of long chains of about 3000–6000 glucose units (hexose) and in the cell wall these units are associated parallel to one another to form microfibrils. Microfibrils in turn consists of smaller sub-units called micelles. About 20 micelles with a cross-sectional area of about 62,500Å form a microfibril.

The spaces among the micellar strands constitute intermicellar spaces. Thus the microfibrillar system of cellulose is interpreted by a system of intermicellar spaces (capillaries) of various sizes. These capillaries are filled with non cellulosic substances and water. Though these capillaries are filled with these materials, other substances can still pass through them, thus rendering the cell wall porous.

In the cell wall, orientation of microfibrils is in various ways. In the primary wall the microfibrils are randomly oriented transverse to the long axis in the first instance subsequently becoming longitudinally inclined as the primary wall starts growing. In the secondary wall, separation of three S layers is mainly based on the orientation of the microfibrils. The helix of microfibrils is lax and nearly horizontal in S_1 and S_3 and steep in S_2 which is the thickest. The primary wall differs from the secondary wall by possessing random orientation of the microfibrils. The chemical composition of these layers also differs.

Lignification: In the secondary wall lignin gets deposited and the cell becomes dead. The impregnation of cell wall with lignin is called lignification. Consequently cells become hard, thick walled and dead (e.g., Sclerenchyma, Xylem and Phloem).

Suberization: Cell walls of cork cells and casparian strips of endodermis get deposited with a layer of suberin by a process known as suberization.

SPECIAL STRUCTURE OF CELL WALLS

Primary Pit fields

The Primary walls of young cells stretch and increase in surface area and thickeness as the cell grows. However, certain areas of the wall remains thin and these portions are termed as primary pitfields or primary pits or primordial pits. A characteristic feature of these primary pit fields is the presence of thin prosoplasmic strands known as 'plasmodesmata'. The plasmodesmata connects the protoplasts of the neighbouring cells and are present in all living cells of higher plants. The living protoplasts are united to form a single unit because of the plasmodesmata. The interconnected protoplast is called 'synplast' and the compartment external to the protoplast the 'apoplast'. The apoplast consists of cell walls, intercellular spaces and the lumen of non-living cells such as tracheids and fibres. Plasmodesmata are easily seen in the endosperm of seeds like *Phoenix, Aesculus* and *Diospyros* and in the cotyledons of some plants. On plasmolysis, the protoplast withdraws from the wall in all places except where plasmodesmata are present. Plasmodesmata occur in groups sometimes more or less evenly distributed over the entire walls. In a variety of cells, branched plasmodesmata are observed, for example between companion cells and sieve elements etc. The branches depart from a median cavity or module to the cells on both sides. There is a possibility that such branches are primary plasmodesmata which become joined by the dissolution of a median portion of the common wall of the neighbouring cells. Plasmodesmata play an important role in the transport of materials and the relay of stimuli. Viruses can pass from one cell to another via the plasmodesmata.

Pits

Certain portions of the cell wall remains thin even after the secondary wall is formed and they consists of only primary wall material. These areas which are of variable shape are called pits. Pits can develop over the primary fit fields and then one or more pits may develop in the pit field or pits may also develop on the primary wall devoid of pit fields. The pits are areas through which substances pass from cell to cell. The concentration of plasmodesmata in living cells in the region of the pit membranes is an additional proof of the pit being a channel of exchange. Each pit has a complementary pit exactly opposite to it in the wall of the neighbouring cell. Such pits form a morphological or functional unit called pit pair. The cavity formed by the break in the secondary wall is called pit cavity. The membrane representing the primary wall and middle lamella that separates the two pit cavities of the pit pair is called the pit membrane or closing membrane. The opening of the pit on the innerside of the cell wall, i.e., on that side facing the lumen of the cell is called the pit aperture.

Two principal types of pits are recognized: Simple pits and Bordered pits. The main characteristic of bordered pit is that the secondary wall develops over the pit cavity to form an over arching roof with a narrow pore in its center. In a simple pit no such development of the secondary wall is present. If the two pairs of pits are simple they are called — a simple pit pair, if the two pits are bordered — a bordered pit pair, if one pit is simple and the other bordered — a half bordered pitpair, if the pit has no complimentary pit in the adjacent cell or if it is opposite to an intercellular space it is termed a blind pit. Sometimes two or more pits are found opposite to one large pit — such an arrangement is called unilateral compound pitting. Simple pits are found in parenchyma cells with thickened walls, libriform fibers and sclereids. Bordered pits are present in tracheary elements and fibre tracheids.

The bordered pit is more complicated than the simple pit in its structure (Fig 3.1) and is variously shaped. In the bordered pit, the cavity that is formed by the over arching of the secondary wall is called pit chamber and the opening in the secondary wall that faces the cell lumen is called the pit aperture. If the secondary wall is very thick a canal the pit canal is formed between the cell lumen and the pit chamber. In the pit canal, two openings are there — that facing the cell lumen is termed the inner aperture and that nearest the pit chamber, the outer aperture. In some bordered pits, the pit membrane is thickened in its central portion which is disc shaped and is termed 'the torus'. The diameter of the torus is wider than that of the pit aperture. When the torus is in the median position, i.e., in the middle of the pit pair, water can easily pass from one tracheid to another. In a pit pair where the torus is in a lateral position, i.e., pressd against one of the pit apertures, the passage of water is prevented. Most of the tori in late wood and all of them in heart wood are always in a lateral position. In this con-

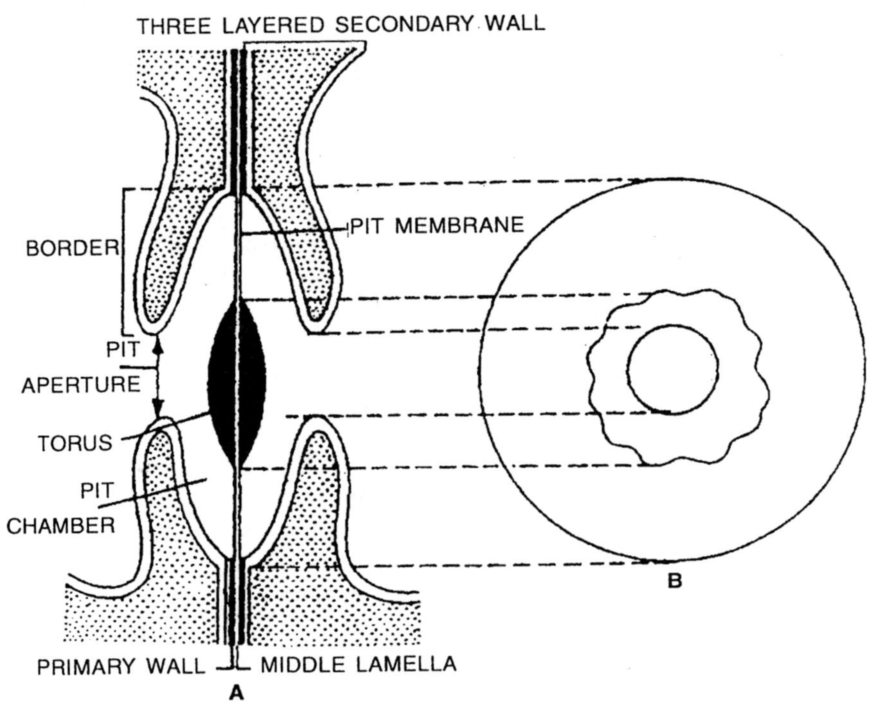

Figure 3.1: Diagrams of bordered pit pairs. (A) Sectional view; (B) Front (surface) view.

dition, the pit is called 'aspirated'. The presence of a torus is characteristic of the bordered pits of Gnetales, *Ginkgo* and most of Coniferales. Tori occur rarely in Ophioglossales and angiosperms. Bordered pits found in tracheary elements vary in shape and arrangement.

Patterns of Bordered Pits

The bordered pits in vessel walls of angiosperms show three main types of arrangement: (1) scalariform, (2) opposite, and (3) alternate.
If the pits are elongated or linear and make a ladder like arrangement, it is called scalariform pitting. When the pits are arranged in horizntal pairs, it is called opposite pitting. If such pits are crowded, their borders assume rectangular outline in face view. When the pits occur in diagonal rows, the arrangement is called the alternate pitting.

CHEMICAL NATURE OF THE CELL WALL

The most common component in plant cell wall is carbohydrate cellulose. It is associated with other compound carbohydrates like non-cellulosic polysaccharides hemicellulose and pectic compounds. The fatty compounds cutin, suberin and waxes also occur in varying amounts in some cells. Various other compounds and mineral substances are also present. Water is common constituent of cell walls.

Cellulose: It is hydrophilic, crystalline compound with chemical formula $(C_6H_{10}O_5)_n$. It is a polysaccharide and closely related to starch. Its molecules are chains or ribbon like structures with 100 or more of the glucose residues held together to O_2 bridges.

Non-cellulosic Polysaccharides: The galactans, the xylans and the arabans which are closely allied to cellulose are commonly found in the cell walls.

Pectic substances: These are derivatives of polygalacturonic acid and occur in 3 general types: Protopectin, pectin and pectic acid. Pectic compounds are amorphous, colloidal substances, plastic and highly hydophilic. The pectic substances constitute the inter cellular substances that binds together the wall of the individual cells.

Gums and Mucilages: These are also compound carbohydrates of the cell walls related to pectic compounds and possess the property of swelling in water. Gums exude as a result of physiological and pathological disturbances that induce break down of walls and cell contents where as the mucilage occur in some gelationous type of cell walls.

Lignins: This is an organic compound of high carbon content, distinct from the carbohydrates. It is found in all the three wall layers i.e., primary wall, middle lamella and secondary wall. The impregnation of lignin starts in the intercellular lamella and spreads to primary and secondary walls. The process of conversion of cellulose into lignin is called lignification.

Cutin, Suberin and Waxes: The cutin and suberin are fatty substances which are not meltable and are insoluble in fat solvents. Cutin and suberin are closely related, highly polymerized compounds consisting of fatty acids. Suberin occurs along with cellulose in the cork cells of periderm, endodermis and exodermis of many plants. The phenomenon of impregnation of walls with suberin is known as suberization. Usually cutin forms a continuous layer, the cuticle, on the surface of epidermis of many plants. It also occurs along with cellulose in the outerwalls of epidermis. The process of impregnation of walls with cutin is called cutinization and the formation of cuticle is 'cuticularisation'.

Waxes are also fatty substances which are meltable and are easily extracted with fat solvents. Waxes are associated with cutin and suberin and are deposited on the surface of the cuticle in various forms. The deposition of wax is responsible for glaucous condition of many fruits, leaves and stems. The deposition of these fatty substances checks the transpiration and protect the leaves from the leaching effects. The hard varnish like cuticle protects against the penetration of potential parasites and also against mechanical injuries.

Mineral substances: Mineral substances like silica, calcium carbonate, calcium oxalate and several organic compounds like tannins, resins, volatile oils also may impregnate the cell walls. Silica or sand particles are common in the walls of grasses and horse tails (*Equisetum* sps). Calcium carborate is deposited in the form of cystoliths in *Ficus* sps. and others. Organic compounds are frequently deposited in the xylem walls when it changes from sapwood to heart wood.

INTER CELLULAR SUBSTANCES

The meristematic or young cells do not possess any intercellular spaces, but when they become older, their walls split at certain places and give rise to small spaces called intercellular spaces. These are filled with air or water. There are two types of inter-cellular spaces, viz., schizogenous intercellular spaces and lysogenous intercellular spaces.

Schizogenous Intercellular Spaces

Most commonly seen due to separation of cell walls from each other along more or less extended areas of their contact. In such cases, the intercellular substances dissolves partly and an intercellular space develops. Ultimately, it becomes big in size and is called schizogenous cavity. The ordinary intercellular spaces and schizogenous cavities form an intercommunicating system of long, intercellular canals which facilitates the diffusion of gases and liquids from one part of the body to the other. Resin ducts of coniferales is a typical example. The cells lining the cavity are secretory in nature and release their products in the intercellular duct canal.

Lysogenous Intercellular Spaces

These spaces arise through the dissolution of entire cells and hence known as 'Lysogenous Intercellular spaces'. These cavities of intercellular spaces store water, gases and essential oils. Examples are found in water plants and many monocots. The secretory cavities in *Eucalyptus, Citrus, Gossypium* are good examples.

CHAPTER IV

MERISTEMS

In marked contrast to animals, plants, vascular plants particularly, exhibit the open system of growth comprising the formation of organs or the parts of organs and various types of tissues throughout the entire period of their life. In these vascular plants such a type of distinctive growth is governed by or restricted to the existence of a region bearing perpetually dividing cells and the region is confined mostly to the apices of the plant body, i.e., the shoot and the root. Such a region with perpetually dividing cells is called a meristem. This term was coined for the first time by Nageli (1856) to designate a type of dividing tissue comprising parenchymatous cells. In older regions such a type of meristem is restricted to vascular cambium, cork cambium and the shoot and root apex. The term meristem has been derived from Greek word *Meristos* which means divisible.

The meristematic cells have definite characteristic features; they are isodiametric in shape, compactly placed, having thin cellulose wall, vacuole may or may not be present, richly cytoplasmic and large nucleated. The plastids may be absent and if present they are in the proplastid stage. However, the meristematic cells of the cambium show certain features deviating from normal meristematic cells. These cells are fusiform in shape, bear thick cell wall, may contain plastids and also ergastic substances such as starch and tannin. The meristematic cells retain the property of division and continue to divide to form derivative cells and as such these cells are formed as initiating cells. The derivative cells gradually loose their meristematic property and become differentiated into mature cells and this process is called differentiation. The mature cells formed lose the property of division and living contents sometimes become permanent.

CLASSIFICATION OF MERISTEMS

I. Classification based on origin and development:

On the basis of origin and development of initiating cells, three categories are recognised.

1. Promeristem or primordial meristem: The term promeristem is used to refer to a group of earliest and youngest meristematic cells of a growing organ. Promeristem further divides and forms primary meristem. It occupies a small area at the tip of the stem and root.
2. Primary meristem: These are derived from promeristem and found at the tips of root, stem and appendages. The cells divide in all possible planes and form the fundamental part or primary structure of a plant.
3. Secondary meristem: It is derived from primary permanent tissues which have the capacity of division. These are always lateral in position and develop either at the time of emergancy to heal the wound or effect secondary growth or to form cork.

II. Classification on the basis of position:

On the basis of position in plant body meristems are classified into:

1. Apical meristems: They are found in the terminal parts of the growing shoots and roots.
2. Intercalary meristems: They are found intercalated between permanent tissues. They are found at the base of the leaf in *Pinus*, at the base of internode in grasses and *Equisetum* and below the node in *Mentha*.
3. Lateral meristems: These are arranged parallel to the sides of organs and normally divide periclinally or radially and give rise to secondary permanent tissues.

III. Classification on the basis of function:

Haberlandt considering the nature of function of meristem recognised three important categories:

1. Protoderm meristem: It is the outermost layer of the young growing region which develops to form epidermal tissue system.
2. Procambium meristem: It is composed of narrow, elongated, prosenchymatous, meristematic cells that gives rise to the vascular tissue system.
3. Ground meristem: It is composed of large, thin walled cells which develop to form ground tissue system, i.e., cortex and pith.

IV. Classification on the basis of Plane of division

On the basis of pattern and plane of division of meristematic tissues, meristems are divided into:

1. Mass meristem or block meristem: Cell divisions in these meristems occur in all planes so that the plant body shows isodiametric or irregular shape. This type of growth takes place in endosperm and young embryo and also in the formation of spores and sperms.
2. Rib or File meristem: The rib meristem divides by divisions perpendicular to the longitudinal axis of the plant organ resulting in parallel

longitudinal rows of cells in cylindrical plant parts. e.g., cortex and pith of stem and cortex of root.

3. Plate merstem: The plate meristem divides mostly by anticlinal divisions (divisions perpendicular to the surface) resulting in surface growth as in leaf blades of angiosperms and epidermis in particular.

CYTOLOGICAL AND MOLECULAR ASPECTS OF THE MERISTEMS

The meristematic cells are usually thin walled, more isodiametric in shape than the cells of the mature tissues and are relatively richer in protoplasm. But it is not possible to distinguish by morphological criterion from which meristematic cells can be distinguished from unspecialized mature cells. The protoplasts of the meristematic cells are devoid of reserve materials and crystals and the plastids are in the proplastid stage. Apical cells of the most of the plants especially among the angiosperms, the vacuoles are very small and are not obvious and scattered through out the protoplast. But in Pteridophytes and in many Spermatophytes, the apical cells contain conspicuous vacuoles. In general, the larger the meristematic cell the greater the degree of vacuolization. The size of the meristematic cells varies between 10–20 μm. The wall of the meristematic cells is very thin and weak but certain cells in apical meristems have thick wall. The nucleus is always a dominant part of the cell, constituting as much as 50% of the volume. In many cases it is the only organelle that is easily detected by light microscopy, because all the other organelles are so small and few. Mitochondria are present and constitute about 12% of the cytoplasm. Dictyosomes and ER are usually rather sparse and only rarely produce vesicles. The organelles are rapidly growing, dividing, producing ATP, harmones and other metabolites. The apical meristems and their cells are extraordinarily dynamic parts of the plant.

In *Arabidopsis thaliana* (a Cruciferous weed which has smallest genome and fully studied) several genes have been identified which are responsible for proper functioning of meristems. The molecular genetic analysis of meristems result in the identification of some genes such as 'CLAVATA', 'KNOTTED' etc. KNOTTED I in *Arabidopsis* play a role in leaf morphology, development and regulation of shoot apex. This gene is expressed only in certain groups of cells. The mgo I and mgo II genes are responsible for enlarged meristems and produce less leaves and floral organs.

The following genes are found responsible for molecular organization in *Arabidopsis* and *Zea mays*. The site of action is also indicated in the following table.

S. No.	*Name of the gene*	*Site of action*
1.	KNOTTED I (in *Zea mays*)	Specification of shoot- meristem
2.	S T M (Shoot meristem less)	Specification of shoot meristem
3.	WUCHSEL	Maintenance of shoot meristem
4.	CLAVATA I	Regulation of shoot meristem size
5.	AINTEGUMENTA	Regulation of size of lateral organs
6.	L_1 L	Specification of Tunica Layer

APICAL MERISTEMS

In higher Algae, the meristematic activity is restricted to the terminal cell of the filament. Such a cell is called apical cell. Growth by apical cell also takes place in the Bryophytes and Pteridophytes. However in Gymnosperms and Angiosperms growth takes place by a group of initials and their immediate derivatives. The term apical meristem is used to refer to these initials and their immediate derivatives. The terms shoot apex and root apex are used as synonyms of apical meristem. Usually the part above the youngest leaf primordium in a shoot apex is considered as apical meristem.

Theories and Concept of Apical Organization

Wolff (1759) recognised for the first time that the shoot apex was responsible for the growth of plant body. Since then, it has attracted the attention of several workers. Several theories have been put forward from time to time to explain the mode of growth found in apical meristem.

1. Apical cell theory: The theory was first proposed by Hofmiester (1857) and advanced by Nageli (1878). According to this theory, a single apical cell is the structural and functional unit of apical meristem which governs the entire process of apical growth. Such organisation is found only in Cryptogams (Algae, Bryophytes, Pteridophytes, Fig. 4.1 A), The single cell is pyramidal in shape. Typically it divides parallel to its faces (except the outermost) so that the resulting derivatives have an orderly arrangement indicating their ontogentic relation to apical cell.
2. Histogen theory: Hanstein (1868) proposed Histogen theory after studying several Angiosperm shoot apices and embryos. According to this theory, the body of the plant arises from a massive meristem of considerable depth comprising three precursors of tissue regions, the histogens, each beginning with one to several initials at the apex arranged in superposed stories. The histogens are the Dermatogen (precursor of the epidermis), the Periblem (precursor of the cortex), and Plerome (precursor of the central vascular cylinder).

For long time Hanstein's theory found acceptance. However, later research showed that these histogens could not be well established in the shoot apex. There are two objections against this theory: (1) In the higher plants (Spermatophytes) it is not possible to demarcate periblem and plerome, (2) No predetermination of the future tissues can be traced to the initials of the apical meristem.

3. Tunica-Corpus theory: Schmidt (1924) proposed Tunica-Corpus theory based mainly on the planes of cell division in the apical meristem. According to this theory the initial region of the apical meristem consists of: (1) the tunica, one or more peripheral layers of cells which divide in planes perpendicular to the surface of the meristem (anticlinal division), and (2) the corpus, a body of cells several layers deep in which the cells divide in various planes. Thus the corpus is a core of cells that adds bulk to the apical meristem by increase in volume and the tunica is a mantle of one or more cell layers that maintain their continuity over the enlarging core by surface growth.

   Sometimes, especially among monocotyledons, a few periclinal divisions occur in the tunica, which is a contradiction of the original definition of the tunica and therefore Popham and Chan (1950) introduced the term mantle for all the outer layers of apex which can be distinguished histologically from the inner cell mass, the core, without taking into account the planes or division in these layers. They retained the term tunica for those layers in which only anticlinal divisions take place.

4. Anneau initial and meristeme d'attente theory: Buvat (1955) proposed that peripheral and subterminal regions are real initiating zones whereas distal zones are inert in nature. He recognized three distinct regions in apical meristems, (i) the anneau initial (the peripheral active zone), (ii) the meristems d' attente (the waiting meristem which becomes active only during formation of terminal flower or inflorescence) and (iii) the meristeme medullaire (the central pith region). Meristeme d'attente consists of two parts, the promeristeme sporogene and promersteme receptaculair which thus contain part of tunica (except anneau initial) and part of corpus (except meristeme medullaire).

5. Korper-Kappe (Body Cap) theory: Proposed by Schuepp (1917), this theory explains the relationship between cell patterns and growth in root tip. The planes of those divisions that are responsible for the increase in the number of vertical cell files in the meristematic region of the root. In this growth, longitudinal and transverse divisions are combined in such a way that, at the point of doubling of files, transverse and longitudinal walls in a given file form a pattern resembling a T. If the root cap has its own initials, the direction of the top stroke (horizontal

bar) of the T in the cell files sharply differentiates the root cap from the body of the root (Fig. 4.2 C).

Types of Stem Apices

Wolff (1759) discovered that the new leaves and tissues of the stem arise in the very apex of the stem which he termed 'punctum vegetationis'. Today the term shoot apex is generally used. The shoot apex proper is considered as that terminal part of the shoot immediately above the upper most leaf primordium. There are great differences in the shape and size of the shoot apices. In a median longitudinal section it appears generally more or less convex. But in *Anacharis, Elodea. Myriophyllum* and some grasses the shape of the apex is a narrow cone with a rounded tip while in a few plants like *Drimys* and *Hibiscus syriacus* it is slightly concave. Before the initiation of each leaf the apical meristem widens considerably and after the appearance of leaf primordium it again becomes narrow. This phenomenon is rhythmic, i.e., it recurs with the initiation of each leaf or pair of leaves. Schmidt (1924) introduced the terms minimal and maximal areas of the apex. For the period between the successive initiations of two leaves or two pairs of leaves he suggested the use of the term 'plastochron'. The shoot apices of dicots with opposite leaves like *Linociera, Ligustrum, Syringa, Vinca* etc. are suitable for the study of plastochron changes.

In the angiosperms, the shoot apices are usually small. The diameter varies between 90 μm (in certain grasses) and 130–200 μm (in many dicots). In the Banana it is 280 μm, certain Palmae and in *Nymphaea* 500 μm and in *Trichocereus* it is 700–800 μm. They can be as large as 1500 μm, in some Cactaceae. Meristems as large as 3,000 μm wide occur in *Cycas revoluta*. Generally, the difference in the diameter of the apices in gymnosperms are much greater. The apices of the conifers are cone shaped and fairly narrow.

In vascular plants the development in stem apex follows more or less definite pattern and these appear to be a feature of the major groups of plants. These patterns manifest rather increasing complexity from lower to higher groups of plants. In the lower groups the initiating cells constitute a uniseriate superficial layer and there is no clearcut evidence of existence of diferentiated tunica and corpus layers. However in higher plants the initiating cells are arranged in two layers. The superficial layer of cells form the tunica which is usually uniseriate with its cells dividing anticlinally and the lower layer of cells gives rise to the corpus. On the nature of the formation of the tunica and the corpus the stem apices may be classified into the following groups.

1. The primitive stem apices: In this group, the apices where simple surface initials on division form no distinct tunica and corpus layers, e.g., species of *Selaginella, Isoetes*, Cycads etc. *Lycopodium* stands as a very

good example representing this type of stem apex. In this, the initiating layer is weakly defined and the uniseriate surface layer freely divides anticlinally and periclinally, as no morphological differentiation between the cells can recognised. The anticlinal divisions increase the surface area and the periclinal divisions increase the inner core of cells.

2. Stem apices with ill defined tunica corpus: The distinction between tunica and corpus perhaps starts with some of conifers. In Pinaceae like *Pinus* and *Abies* the initials form a terminal uniseriate group of cells. Both anticlinal and periclinal divisions take place in the formation uniseriate superficial envelope over a central mass of cells. The former suggests, the existence of the tunica in appearance but its periclinal divisions in parts of apex makes it difficult in drawing a line between tunica and corpus. But some of the higher Gymnosperms show a special structure. In Pinaceae the initials or a small group of cells arrange in a single tier with divisions both anticlinal and periclinal. Soon a dermatogen like layer is organised covering the central mass of cells. The fundamental difference from the Pinaceae is that when the dermatogen like layer is organised the cells only have anticlinal divisions with occasional periclinal divisions. The central cells always show periclinal or irregular divisions. In this condition the demarcation between tunica and corpus is easier than in Pinaceae. In certain Gymnosperms like *Ephedra, Araucaria* etc, the tunica is uniseriate and the corpus forms the bulk of cells.
3. Stem apex with distinct tunica and corpus: When compared to lower plants the angiosperms show definite and distinct apical meristem zone. The two sets of initials are placed one above the other which give rise to tunica and corpus. The tunica as we know is very rarely periclinal and ranges from one to few cells in thickness and 2 or 3 layers probably representing the common feature. A single layer of tunica as observed in *Avena* and *Triticum* probably represents a highly specialised condition. But in certain cases like *Zea* occasional periclinal divisions are recorded in tunica. The limits of the two groups of initials is often definite although in certain exceptional cases like members of Cactaceae, *Sinoealanus, Vinca rosea*, such distinction can not be drawn out and the apices resemble those of Gymnosperms showing a weak differentiation of tunica and corpus.
4. Haberlandt (1914) recognised three distinct primary meristimatic zones namely protoderm forming epidermis, hypodermis, the ground meristem forming the medulla, pith and procambium forming vascular elements.
5. Popham (1951, 1957) recognises 4 to 3 zones in the shoot meristem:

(a) Surface zone or mantle composed of a few to several layers corresponding to tunica.

(b) A zone of subapical mother cells which are highly vacuolated and less active in their cell divisions.
(c) Central region of cells which will give rise to pith.
(d) Peripheral zone which will give rise to cortex and procambium. In certain plants he was able to recognise a fifth zone just below the subapical region of mother cells with the cells stretched across the axis. This is the zone resembling that of the cambial zone.

Significance of Tunica Corpus theory: This theory has well served to understand the nature, form and pattern of stem tips of seed plants. In view of these features of meristem, the position, number and behavior of initiating cells in the stems of seed plants is now in some measure known. Although this theory does not throw much light upon morphological significance of these two layers it has got the topographical value in the detailed study of development when the distinctiveness of tunica and corpus is known. The regions are constant in structure and function showing marked variability with seasonal conditions, vigour of growth, age, position of plant and morphology of stem tip concerned. In recent years accounts also have been made in the light of tunica and corpus to aim at phylogeny of plants. In this regard Griffard's work is very much appreciated.

Newman's Classification of Shoot Apices

Newman (1965) used the term continuous meristematic residue (*cmr*) to refer to a group of permanent initials in the shoot apex, and on the basis of form of the *cmr*, shoot apices are divided into the following types.

(i) Monoplex type is the characteristic of ferns, in which *cmr* is in the superficial layer only. Any one cell contributes to growth in both length and breadth.
(ii) Simplex type is common in gymnosperms. The *cmr* is restricted to a single layer. Cells are undergoing divisions both periclinally and anticlinally. Both the divisions are necessary for bulk growth.
(iii) Duplex type is common in angiosperms. The *cmr* is present in two layers, the first layer undergoes anticlinal division and second layer periclinal divisions. Thus, two contrasted mode of growth is evident, hence these are known as duplex type of shoot apex.

Cytohistological Zonation in the Promeristem of Vegetative Shoot Apices

Shoot Apex of Pteridophytes

In this group there are one or more initials which can be easily distinguished from the neighbouring cells. These initials give rise to all the cells of the apex.

If only one initial is present it is termed the apical cell and if more than one cell they are termed 'apical initials'. The single apical cell is usually tetrahedral in shape. The single apical cell is found in Psilotales, *Equisetum* and in some other ferns. The apical cells are generally 4-sided, but in some water ferns like *Salvinia* and *Azolla*, sometimes in *Selaginella* they are 3-sided. The ferns (Filicinae) with a single apical cell are considered to be evolutionarily more advanced than those with several apical initials.

Shoot Apex of Gymnosperms

Since 1937 much research has been made on the structure of shoot apex in gymnosperms. The characteristic feature of all the gymnosperms is the direction of the cell division in the surface of the apex which is both anticlinal and periclinal. So, this layer represents the initiation zone of the entire apex and has been termed as the 'surface meristem'. The striking feature of the most gymnosperm apices is the occurrence of a distinct zone of central mother cells. The cells of this zone are characterized by their size, numerous large vacuoles and thick walls. Along the sides and base of the central mother cell zone, the other apical regions develop as a result of the diagonal and horizontal divisions of the central mother cells. In this way the peripheral zone or flank meristem is laterally developed and the rib meristem zone (central meristem) from the base. The term 'rib meristem' was introduced by Schuepp (1926) to describe that type of meristematic issue which shows vertical series of transversely dividing cells. According to Popham (1952), three principal types of gymnosperms can be distinguished on the basis of the structure of the shoot apex — the *Cycas* type, the *Ginkgo* type and the *Cryptomeria* — *Abies* type.

The *Cycas* type (Fig. 4.1 C): In this type three meristematic zones can be distinguished namely: (a) the surface meristem, in which the cells divide anticlinally, periclinally and diagonally. The cells of this zone give rise to the epidermis and the other apical meristmatic zones. (b) the rib meristem which is situated in the central region of the apex below the surface layer. These cells also divide periclinally, anticlinally and diagonally. (c) The flank meristem which enlarges by cell division within the zone itself and by the addition of cells from the surface layer and from the periphery of the rib meristem. This gives rise to cortex, the procambium and the leaf primordia.

The *Ginkgo* type (Fig. 4.1 D): In this five meristematic zones are distinguishable. They are: (a) the surface meristem which consists of the apical initials, (b) the zone of central mother cells which occurs in a median position below the surface layer. The cells are large, polyhedral and are arranged irregularly. The divisions in these cells occur in various planes, (c) the cambium like 'transitional zone' which is cup shaped. This zone forms a transitional zone between the central mother cells and the zones below

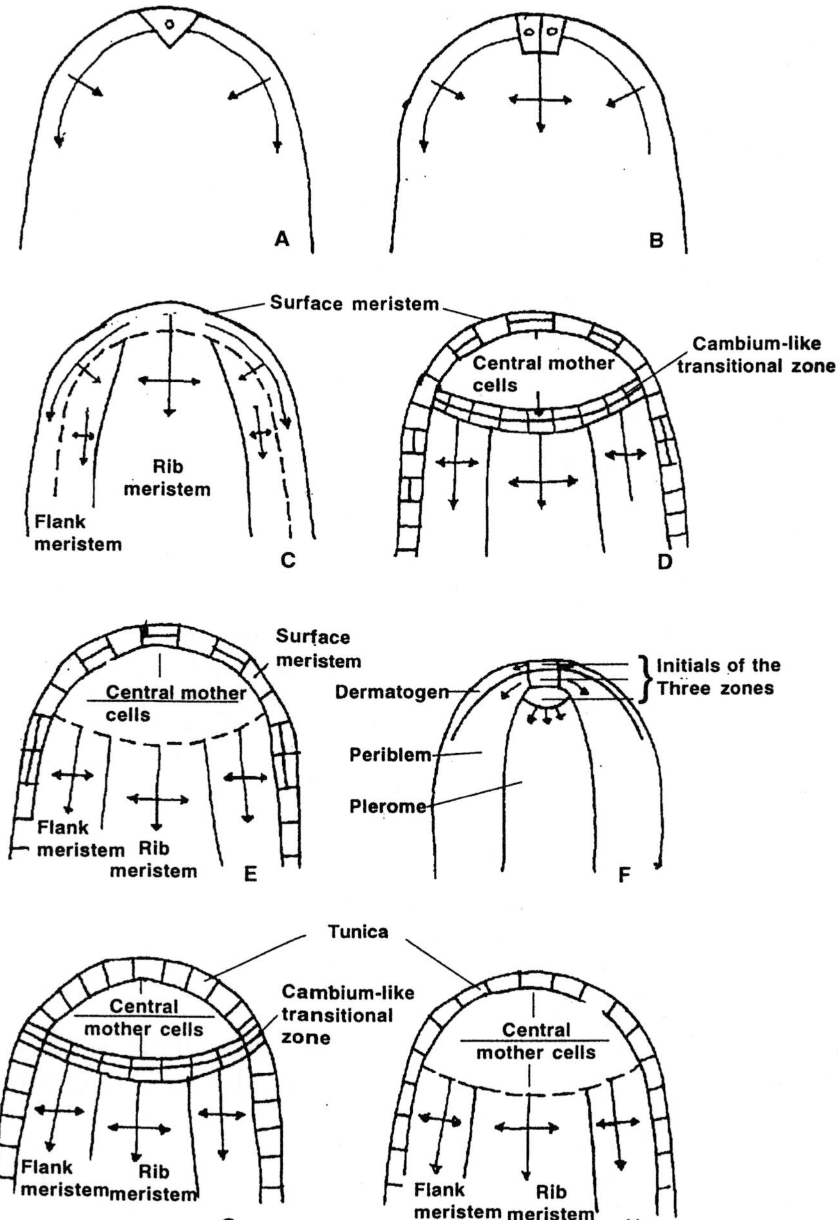

Figure 4.1: Diagrams of cytohistological zonation in the promeristem of vegetative shoot apices: (A) Pteridophyte with single apical cell; (B) *Selaginella* type with 2–5 apical cells; (C) *Cycas* type; (D) *Ginkgo* type; (E) *Cryptomeria-Abies* type; (F) Schematic representation of the histogen of angiosperms; (G) *Opuntia* type; (H) Usual angiosperm type (after Fahn, 1982).

them, (d) the rib meristem is under the central portion of the cambium like transitional zone, (e) the flank meristem which forms a cylinder surrounding the rib meristem and this is a continuation of the cambium like transitional zone (Fig 4.1 D). Other examples are: *Zamia, Sequoia sempervirens, Microcycas* and *Pseudotsuga*.

The *Cryptomeria* and *Abies* type (Fig 4.1 E): In this four meristematic zones can be recognizable. This type is almost similar to the *Ginkgo* type, but the cambium like transitional zone is absent here, e.g., *Pinus, Taxus, Ephedra* etc.

Shoot Apex of Angiosperms

Hanstein (1868) has put forward the Histogen theory. According to Hanstein, three zones can be distinguished in the shoot apex of angiosperms, viz., (a) an outer most zone, the 'dermatogen', a central zone 'plerome' consists of irregularly arranged cells and a hollow cylindrical zone of several layers of cells between the dermatrogen and plerome- the 'periblem'. Hanstein stated that these zones develop from independent groups of initials, which act as direct histogens. According to this theory, the meristems are destined from the beginning to produce certain tissues, i.e., the epidermis from the dermatogen, the cortex and the internal tissues of the leaf from the periblem, and the central cylinder from the plerome. This theory is accepted for a long time, but later researches, disproved it, because of the following. (1) In most Spermatophytes, it is not possible to distinguish clearly between the periblem and plerome, (2) No predetermination of the mature tissues can be traced in the various initials.

In 1924, the theory of Schmidt, which divides the apex into two regions — the 'tunica' and the 'corpus' was postulated. According to this theory, no constant relationship can be traced between the particular initials of the promeristem and the inner tissues of the shoot. The two regions recognized by this theory are usually distinguished by the planes of cell divisionis in them. The tunica consists of the outer most layer or layers of cells which surround the inner cell mass the corpus. The plane of cell division in tunica is principally anticlinal. In the corpus, the planes of cell divisioni is in all directions. The tunica enlarges in surface area and the corpus in volume. The tunica-corpus theory is very adaptable one and till today, it is generally accepted in literature.

As already mentioned, the tunica consists of a single layer or few layers of cells which surround the inner meristem. The number of layers is not always constant in a given genus or family or not even in a species. The number of layers varies in a single plant during different stages of development of the vegetative apex. The number of layers may be 1–9. Single layered tunica (monostratose) is the most common structure, followed by two to four (multistratose), but sometimes as many as 18 layers occur in

Xanthorrhoea media (Staff, 1968). Sometimes, especially in monocots, the periclinal divisions occur in the inner layers of the tunica [if it is multistratose, but the outer most layers have only anticlinal divisions] which is a contradiction of the original definition of the tunica and therefore, Popham and Chan (1950) introduced the term 'mantle' for all the outer layers of the apex which can distinguished histologically from the inner cell mass without taking into account the planes of cell division in these layers. They retained the term tunica for those layers in which only anticlinal cell divisions takes place. The term tunica is used in the broad sense and is equivalent to the mantle of Popham and Chan.

Although the plane of cell division is generally the same through out the entire tunica, cytologically two zones can be distinguished in it viz., (1) Central apical zone consisting of one or few initials which are larger and have larger nuclei and vacuoles than the other tunica cells. These cells take less stain, (2) It is that region on the sides of the apex between the initials and the leaf primordia. It consists of smaller, more darkly staining cells which divide more frequently and among which periclinal divisions may occur close to the primordia.

The corpus is far less homogeneous than the tunica. Among the angiosperms two main types of corpus are recognized on the basis of internal arrangement (Popham, 1952).

The Usual Angiosperm Type

In this three main zones can be distinguished in the corpus (Fig 4.1 N) viz., (a) the zone of central mother cells which probably represent the corpus initials and located below the apical portion of the tunica, i.e., below the tunica initials (b) the rib meristem and (c) the flank meristem. The latter two zones appear as continuations of the central mother cells.

Opuntia Type

In this type, in addition to the zones found in the above type, a cambium like transition zone can be distinguished (Fig 4.1 G). This zone which is cup shaped is found between the central mother cells and the rib and flank meristems. The cambium like transitional zone differs from the other zones of the apical meristems in its height and diameter which vary considerably during the plastochron, reaching its maximum development close to the developing primordium. According to Philipson (1954), this zone is only temporary and it disappears towards the end of plastochron. Examples: *Phoenix dactylifera, Chrysanthemum, Opuntia cylindrica* and *Xanthium pennsylvanicum.*

The cells of the flank meristem are usually derived from the initials of the corpus, but in some plants, they also arise from the tunica initials. The rib meristem consists of cells arranged in rows which narrow towards the apex, they may some times be irregularly arranged. The majority of the divisions in this zone are horizontal, but diagonal divisions also occur. It can be concluded that the epidermis and its derivatives orginate from the outer most layer of the tunica. The flank meristem contributes to the development of the leaf primordia, the cortex, all or part of the procambium and sometimes also the outer region of the pith, and the rib meristem the reminder.

Zonation in Root Tip

Like the shoot, the apical meristem of the root also displays many patterns of growth and there exists no constant relationship between structure of initiating division and delimilation of primary tissues of the axis. In the root unlike the stem, the apical meristem producess the root cap but even away from it since it initiates the root cap and there by occupying subterminal position. The apical meristem of the root differs from that of the shoot in that it forms no appendages nodes and internodes. The root branches are usually initiated endogenously faraway from the meristematic region. Besides these differences the root apex shows no change in shape and structure as seen in stems because of the absence of the leaves.

According to various investigators it appears that there is a single central initial (Fig. 4.2 A, C) or but a few initials in the root apex of spermatophyta. Other investigators such as Clowes (1950), believe that there is a larger group of initials in the median region of the root apex.

Although the histogen theory has been abandoned for the shoot apex, the structure and growth of the root apex are generally interpreted in terms of histogen theory. The apical meristem in the case of root tip is rather short. In number the initials range from one to many and where the initials are more than one, they are arranged either in one row or usually in four rows fairly distinct from one another as uniseriate stripes. In each group there are one to several initiating cells. Further more the initiating cells of the different groups may vary in their diameter and in the rapidity of growth of root. For example in slender roots, the initiating layer is restricted to one, but zonation remains distinct. Where there are more layers of initiating cells the layers invariably lie one above the other along the longitudinal axis of the root.

Each of these groups quickly develops one or more growth zones which are usually marked more clearly than in stem. In many plants these groups of initiating cells represent in strict sense the histogens; although the terms dermatogen, periblem and plerome are not in general view, in the case of stem ontogeny they have been continued to indicate the general zone in the stem and root development. A fourth histogen namely calyptrogen is added

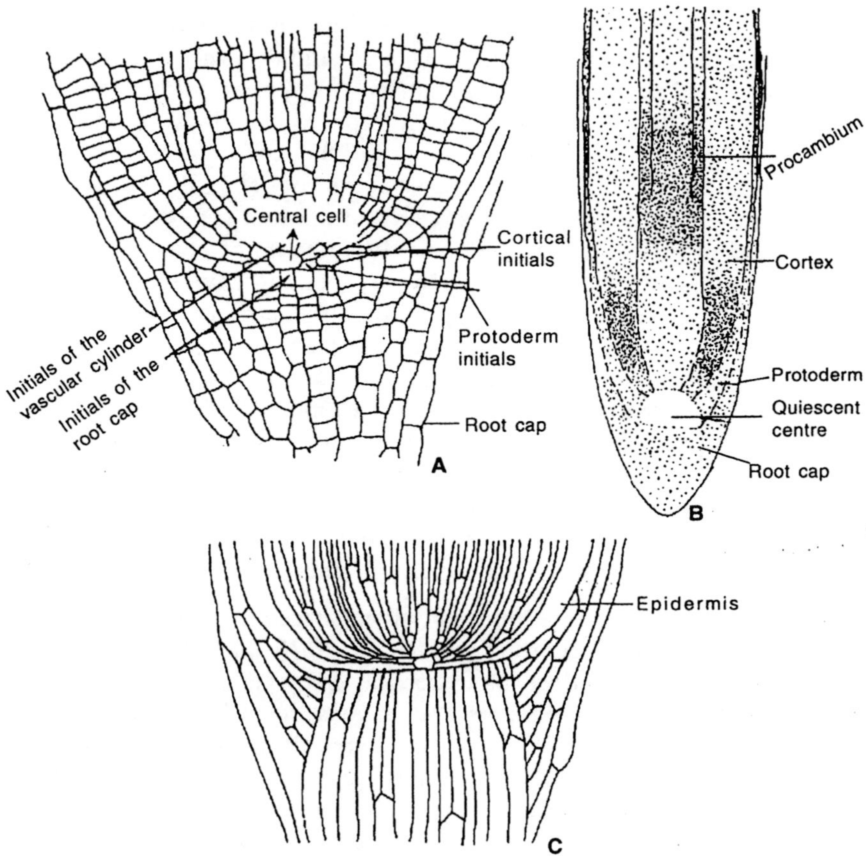

Figure 4.2: (A) Longitudinal section of the root tip of *Triticum vulgare*; (B) Diagram of the root tip of *Allium cepa* showing , by means of shading, the gradation pf mitotic activity in different zones; the most active zone is the most darkly shaded; (C) Diagram illustrating the Korper-Kappe pattern of the root apex of *Zea mays*.

when the root cap has had an independent origin. On the basis of the activity and distinctiveness of histogens, as in stem, the roots are also classified into different types. In the lower vascular plants all the tissues of the root are either derived from a single apical cell e.g., *Equisetum*, members of Polypodiaceae and certain species of *Selaginella* or may be formed from several initials arranged in one tier as in the members of Marattiaceae. All these plants have the same apical structural organisation both in stem and root. This single apical cell or a single group of initiating cells form the entire root and the origin of the root cap from apical cell is clear. In the majority of Gymnosperms there are two groups of apical initiating cells the one forming the central cylinder, the other giving rise to the cortex and root cap. The outermost layer of the cortex differentiating into the epidermis whose origin and growth may be traced to be slightly away from the apical region as a distal proliferation. The outstanding feature in this case is the formation of a zone from the cortex of the root and the formation of endodermis from the innermost layers of the cortex. In most of the Angiosperms the groups of initials occur at the root apex. In Dicotyledons, the terminal group forms the dermatogen and the root cap, the median one develops the cortex while the upper one gives rise to the central cylinder of the root. On this type both dermatogen and root cap have the same origin and the cap may be attributed to as specialised epidermes or dermatogen. In some dicots like Ranalian families, Amentiferae, Ranunculaceae, certain members of Leguminosae and some of monocots, all the parts of the root appear to have arisen from a common initial group. The epidermis eventually differentiating from cortex and the calyptrogen differentiating from lower part and in these forms the zonation in root tip recalls the condition encountered in Marattiaceae. In certain monocots like *Zea*, the outermost initials develop the calyptrogen or root cap, median ones dermatogen and periblem and inner ones plerome which is a condition generally seen in Angiosperms.

Variation in Monocots: In Monocots, all the three groups of root apices found in Dicots are recorded in *Allium sativum, Haemanthus* and *Zephyranthus tubis* respectively However, the first type of root apex meristem is represented by a single layer and is also recorded from *Eucharis, Amaryllis* and *Swevera* and in certain conditions all in *Allium sativum.* Commonly in most of Monocots the outer most layer of initials produces the cap, the median one the dermatogen and periblem and the inner most plerome. In *Agapanthus* and *Crinum*, the root cap, dermatogen and cortex arise from independent set of initials as recorded by Deshpande (1960) which feature was reported long back by Treub (1876). In all these cases where the root cap origin is independent it is termed as a definite meristem called calyptrogen. Rarely in certain members of Monocots like *Pistia* and *Hydrocharis* the origin of Calyptrogen is from an independent meristem which forms both the initiating layers.

Deshpande (1960–61) recognised only three types of root apices in monocots.

1. The root cap organises independently and can not be reorganised into columella and peripheral zone. This condition has been seen in certain members like *Polygonatum oppositifolium*, *Sanseviera thyrsifolia* and *Ruscus hypophyllum.*
2. The cap is recognised into columella and peripheral zone both representing independent histogens. In this condition the peripheral region of the root cap may be independent of dermatogen and the rest of the tissues of the root tip originates from separate initials as is found in

case at *Haemanthus coccinium*. In *Agapanthus exerinum* part of the root is organised from the dermatogen and peripheral layer may be free. The root cap is not free from other tissues of the root apex and there is no distinction of columella and peripheral layers.

Clowes (1956) while studying root tip of *Zea mays* recognises a cap like structure or region placed in between the root cap and active meristematic region and he called this region the quiescent centre (Fig. 4.2 B).

Likewise Deshpande has also recognised such type of quiescent centre in root apices of some members of Tiliaceae and Amaryllidaceae. Although the function of the quiescent layer is not thoroughly understood, Clowes is of the opinion that this centre becomes active at the time of formation of lateral roots or when root tip is injured. Recently it has been observed by Jensen (1966) that this quiescent centre acts as a reserve of zonal cells, which are less sensitive to damage and becomes active while the other part of the root apex is inactive. According to Deshpande the meristematic cells below the quiescent centre form the root cap, while the initiating cells above it form the other tissues of the root.

Evolutionary significance: From a detailed study of the structural organisation of the apices in several of the plants belonging to diversified affinities, it has been now visualised that the evolutionary progress is perhaps from a single centre of growth to a group of three of four centres with progressive increase of independence of zones. From the foregoing description of different types of root apices the origin of the root cap varies. In the lower plants, the root cap has a common origin with all the other parts of the root. In Gymnosperms it is a special distally developed part of the periblem and the formation of the dermatogen in this condition is a bit delayed and is also gone up away from the root cap. In the dicots the root cap appears to be more specialised in its development from the epidermis. In monocots although they show the same pattern of development of root cap as in dicots but in certain exceptional case the root cap appears to be more independent in being derived from a separate meristem as against the case seen in dicots and majority of monocots.

CHAPTER V

SIMPLE TISSUES

Tissue is a group of cells with common origin, structure and function. From the standpoint of number of kinds of cells making up a tissue, tissues are of two types — simple or complex. Simple tissues are homogenous in nature and composed of structurally and functionally similar cells. Parenchyma, collenchyma and sclerenchyma are the common simple tissues. Complex tissues are heterogeneous in nature, being composed of different types of cells performing diverse functions. Xylem and phloem are complex tissues.

All the tissues of a plant which perform the same general function, regardless of position or continuity in the body, may be considered to form, together a tissue system. The cells have physiological similarity or topographical continuity. Epidermal tissue system, ground tissue system or fundamental tissue system and vascular tissue system are the three tissue systems in vascular plants.

PARENCHYMA

In the fundamental classification of plant cells, a cell with a thin primary wall is called 'parenchyma'. In the parenchyma, there is a lot of diversity in cell size, shape, metabolism and function. In herbs, almost all the plant body is made up of one parenchyma or the other; only small portion of collenchyma and sclerenchyma are present.In shrubs and trees, large parts of the plant body consists of parenchyma.

Parenchyma cells are the oldest type of Eucaryotic cells and are the first to evolve. The entire body of alga consists of only parenchyma. When the plants moved on to the land and became upright, they required special support, then collenchyma and sclerenchyma were evolved as modifications of parenchyma. In Algae, Mosses, Liverworts and lower vascular plants, all the parenchyma is involved in photosynthesis. But in land plants division of labour became more sophisticated and various regions became specialised. Some of the generalists became specialists and new type of metabolism arose. Resin canals lined with secretary parenchyma cells became common in Gymnosperms. In Angiosperms even more diverse type of parenchyma cells developed like mucilage cells, gum cells, laticifers, cells capable of rapid movement in plants and highly pigmented cells of flowers and fruits. Parenchyma is the most common tissue, which is morphologically and physiologically simple and unspecialized.

Shape: Parenchyma cells vary in form, but usually isodiametric. However parenchyma cells may also be elongated and merge with prosenchyma cells. The term prosenchyma was earlier used for elongated cells with tapering ends but in view of the fact that different types of cells may be elongated, the term has gone out of use. Parenchyma cells are lobed in spongy mesophyll and palisade parenchyma of *Lilium*. In the mesophyll of Xanthorrhoeaceae the parenchyma cells have folds or projections. In relatively homogeneous parenchyma, the number of facets approaches fourteen with 8 hexagonal and 6 quadrilateral faces (tetrakaidecahedron).

Parenchyma cells with inner wall protuberances occur in various anatomical structures concerned with short distance transfer of solutes such as in nectaries, salt glands and vascular parenchyma. These cells are termed transfer cells.

Arrangement: Mature parenchyma tissue may be tightly packed and without intercellular spaces or may have a well-developed system of inter cellular spaces. For example, the parenchyma of the endosperm of most seeds is devoid, or almost so, of intercellular spaces while in the stems and leaves of hydrophytes the intercellular spaces reach maximal development. The tissue with the prominent intercellular spaces occurring in plants, growing in water-logged soils and aquatic habitats is called aerenchyma. The tissue is considered to be important in aerating the plant. The development of intercellular spaces is either schizogenous or lysigenous. The resin ducts of the Coniferae, the secreatary ducts of the Asteraceaeae and Apiaceae are formed schizogenously. Lysigenous intercellular spaces are formed by the disintegration of entire cells, e.g., water plants.

Cell wall: The cell walls in active vegetative ground parenchyma are relatively thin. They are composed of cellulose, calcium pectate and hemicllulose. The middle lamella may or may not be distinguishable. Plasmodesmata are common in such walls. The microfibrils are variously oriented in the cell wall.

Sometimes the primary wall becomes very thick due to the deposition of hemicellulose, e.g., endosperm of *Phoenix dactylifera*, *Diospyros*, *Coffea arabica* and *Asparagus*. The walls of these cells gradually become thinner during germination. Parenchyma cells with relatively thick and lignified secondary walls are common, especially in the secondary xylem.

Cell contents: The internal structure of the parenchyma cell varies according to its function. Parenchyma cells, which take part in photosynthesis, contain chloroplasts and then the tissue they form is termed chlorenchyma. Certain parenchyma cells contain leucoplasts. Photosynthesizing cells are usually conspicuously vacuolated. Secretary types of parenchyma cells often have dense protoplasts. Parenchyma cells may serve to store different reserve materials which may be found in solution in the vacuoles or in the form of solid particles or liquid in the cytoplasm. Sugars or other soluble carbohydrates and nitrogenous substances may be found in the cell sap. Starch, protein, oils and fats occur in the cytoplasm in the form of small particles. Starch is the most common reserve material in plants and it is found in the endosperm, cotyledons, tubers, fruits, xylem and phloem parenchyma, the cortex etc. Protein and starch grains are found in the cotyledons of many species of the Fabaceae and protein and oil are found in the endosperm of castor and the cotyledons of soybean. In the parenchyma cells of the potato tuber amides and proteins are found in the cell sap and starch in the cytoplasm. Parenchyma cells in flowers and fruits often contain chromoplasts.

Parenchyma is phylogenetically the primitive tissue. Parenchyma can be subdivided into 5 classes based on function: synthetic parenchyma, structural parenchyma, boundary parenchyma, transport parenchyma and storage parenchyma.

Synthetic parenchyma: This is also known as 'photosynthetic parenchyma'. In this, the parenchyma cells often become rich in chlorophyll and is known as chlorenchyma, which helps in photosynthesis by converting light energy into chemical energy and store the energy in carbohydrates.

Meristematic parenchyma: All the meristematic cells are parenchyma cells and these are most active of all synthetic cells. They absorb sugars, water and inorganic nutrients and use them to synthesize whole cells.

Secretory parenchyma: These parenchyma cells secrete large quantities of substances which can move to the exterior, or to a cavity or to duct inside. The secreted material can pass from cell to cell to the site of deposition.

Structural parenchyma — Aerenchyma: This is well developed in the spongy mesophyll of leaves, submerged stems and petioles of aquatic plants like *Typha, Nymphaea, Eicchornia, Juncus* etc.

Boundary parenchyma: This is present in the interfaces between two regions of a plant or between a plant and environment. Epidermis and endodermis are two examples of this category.

Epidermis: It is the boundary between the plant and the environment. Both plant and environment are very diverse as is the epidermis. On shoots, the function of epidermis is the retention of water inside the plant. For this the outer wall becomes less permeable to water due to deposition of waxes. Epidermis can be considered secretory parenchyma because it secretes certain water proofing substances. Epidermis is the outermost layer exposed to wind and animals hence must be resistant to tearing and puncture. The epidermal cells are attached to each other very firmly and are strongly interconnected and can be pulled off as a large sheet. Epidermal cells, therfore, can be classified as structural parenchyma.

Endodermis: The conducting tissues are isolated from the cortex by a boundary of cells which constitute the endodermis. The endodermal cells form a cylinder around the vascular tissues and like epidermal cells abut each other very tightly with no intercellular spaces

Transport parenchyma or Transfer cells: In many parts in plants, the material must be transferred rapidly and in larger quantities over a short distance into a gland, into and out of sieve elements, into a developing embryo. For this purpose, transfer cells are present. The cell walls of these transfer cells are highly modified, having many ridges and papillae on their inner faces forming a labyrinth hence called labyrinthine walls or transfer walls. The labyrinthine walls increase the area of absorption when compared to flat walls.

Sieve elements: In the phloem, sugars are conducted over long distances by specialised parenchyma cells called sieve elements. They have a functional structure in which the cells are elongate and have large holes to allow easy passage of phloem sap.

Light transmission: An unusual type of transport parenchyma occurs in *Frithia*, *Lithops* and other members of Aizoaceae — the window leaf plants of the African deserts. The whole plant is subterranean with only the leaf tips reaching the soil surface, so photosynthesis must take place below ground. To accomplish this, a leaf with a transparent channel acts as an optical fibre to carry light from the surace to the photosynthetic tissue below. The plant remains in cooler, more moist subterranean environment, so that water storage and conservation are maximized.

Storage parenchyma: Most of the plant parts used for food — seeds, fruits, tubers are composed of storage of parenchyma. In these organs, the parenchyma store starches (cereals, potatoes), proteins (beans, peas), or oils (sunflower, safflower, castor). Cells store their storage product fully and even the vacuole and other organelleles are obscured. The storage parenchyma also store amides, tannins, mineral substances, enzymes, mucilaginous substances. Succulent plants, such as cacti and Euphorbias are composed of the parenchyma cells which store water. Such cells are large and the bulk of their volume consists of greatly expanded vacuole surrounded by a thin layer of cytoplasm. The cells are much bigger than those that store starch or protein. A less common means of storage is by increasing the amount of hemicellulose in the wall. These are deposited to such an extent that the walls become thick. Later the hemicelluloses are digested out of the walls and the walls become quite thin as in the seeds o *Coffea, Diospyros* and *Phoenix*.

Functions

1. Parenchyma cells are centres of various physiological activities like photosynthesis, assimilation, respiration, storage, secretion etc.
2. In succulents plants, these help in water storage .
3. In aquatic plants they store air and provide buyonancy.
4. Parenchyma cells of the cambium form secondary tissues.
5. They are helpful in wound healing.

Idioblast: A cell in a tissue that markedly differs in form, size or contents from other cells in the same tissue. Specialized parenchyma cells which produce and store tannins, myrosin (e.g., Brassicaceae, Capparaceae), oil substances (e.g., Lauraceae), mucilaginous substances (e.g., many Monocotyledons, Malvaceae, Cactaceae), resinous substances (e.g., Meliaceae, Rutaceae, Rubiaceae) and crystals of calcium oxalate e.g., Moraceae. These cells significantly differ in size, structure or contents. These specialized cells are known as idioblasts (Gr.: Idio = peculiar).

COLLENCHYMA

Collenchyma consists of thick-walled cells and is regarded as a supporting tissue. The supporting tissues of the plant, i.e., the collenchyma and the sclerenchyma are designated from the functional point of view by the term stereome (Haberlandt, 1918). Collenchyma consists of living, more or less elongated cells which, generally, have unevenly thickened walls, Collenchma differs from sclerenchyma in wall structure and condition of protoplast. Collenchyma has relatively soft, pliable, nonlignified primary walls, whereas sclerenchyma has hard, more or less rigid, secondary walls, which are commonly lignified. Many sclerenchyma cells lack protoplasts at maturity while collenchyma cells retain active protoplasts and can resume meristematic activity. There is a physiological and morphological relationship between collenchyma and parenchyma, and, in places where the two tissues occur side by side, transitional forms can be found between typical collenchyma and typical parenchyma.

Cell shape: The size and shape of the collenchyma cells varies. The cells may be short prisms and resemble neighbouring parenchyma cells or long and fibre-like with tapered ends, but all intermediate shapes and sizes occur. The longest collenchyma cells are found in the central portions of the strands of collenchyma, and the shorter ones on the periphery.

According to the type of wall thickening three types of collenchyma are recognised (Muller, 1890).

(a) Angular collenchyma (Fig. 5.1 A, E): The thickenings are localized in the corners or angles of the cells. Such angular collenchyma is found in the petioles of *Vitis*, *Begonia*, *Cucurbita*, *Morus* and stems of *Solanum tuberosum* and *Atropa belladonna*.

(b) Lamellar or plate collenchyma (Fig. 5.1 B): The thickenings are mainly on the tangential walls of the cells. Such lamellar collenchyma is found in the stem cortex of *Sambucus nigra* and *Rhamnus* and petiole of *Cochlearia armoracia*

(c) Lacunar or Tubular collenchyma (Fig. 5.1 D): The thickenings appear in those parts of the cell wall which face inter cellular spaces. Such collenchyma can be seen in the petioles of the species of Asteraceae, *Salvia*, *Asclepias*, *Malva* and *Althaea*.

Duchaigne (1955) distinguished an additional type, i.e., annular collenchyma (Fig. 5.1 C), in which the cell lumen is circular, or almost round in cross section.

Cell wall: Walls of collenchyma cells consist of cellulose, large amounts of pectin and hemicellulose but no lignin. Since the pectic substances are hydrophilic, collenchyma walls are rich in water. Chafe (1970) has observed that the orientation of the cellulose microfibrils in the successive lamellae of the wall is alternately transverse and longitudinal.

Distribution: The tissue is located either directly beneath the epidermis or one or few layers removed from it. In stems, collenchyma frequently forms a continuous layer around the circumference of the axis as in *Helianthus* stem. Sometimes it occurs in strands, often with in externally visible ridges (ribs) found in many herbaceous stems and those woody stems that have not yet undergone secondary growth e.g., *Cucurbita*. In the petioles, the distribution of collenchyma shows patterns similar to those in the stems. In the leaf blade it occurs in the ribs accompanying the larger vascular bundles (veins), sometimes on both sides of the rib, sometimes on one side only, usually on the lower. Collenchyma is generally absent in underground stems, roots, stems and leaves of monocots.

Functions

1. Provide support to leaves and stems.
2. Protect vascular bundles of leaves by forming cap or bundle sheath.
3. Provide tensile strength for growing organs.

SCLERENCHYMA

Sclerenchyma is a simple but hard tissue composed of cells with thickened secondary cell walls, usually with lignified walls. The word sclerous itself means hard. The cells are mostly nonliving at maturity.

Sclerenchyma cells may differ in shape, structure, origin and development. Generally, sclerenchyma is divided into fibres and sclereides. Fibre is

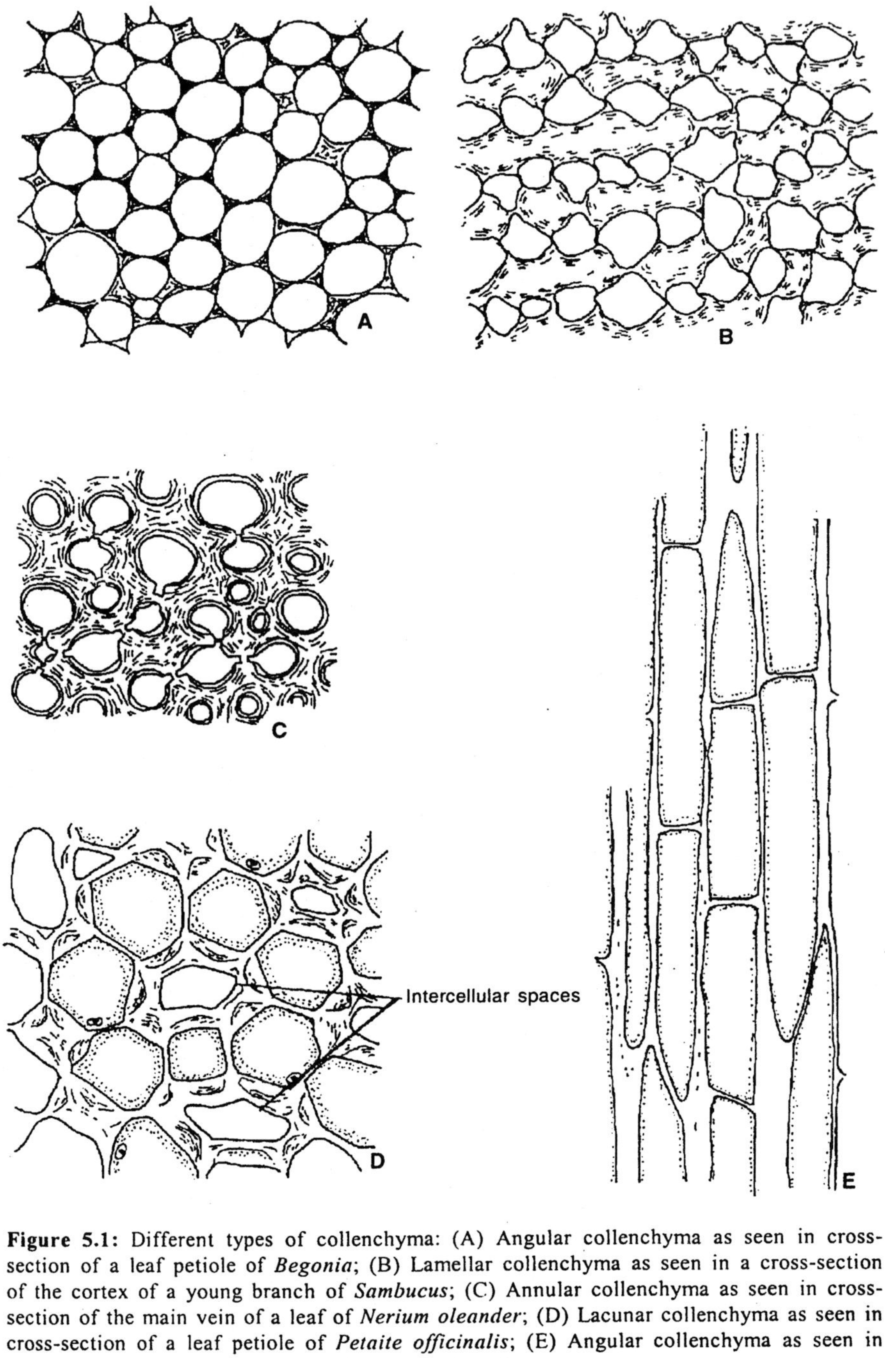

Figure 5.1: Different types of collenchyma: (A) Angular collenchyma as seen in cross-section of a leaf petiole of *Begonia*; (B) Lamellar collenchyma as seen in a cross-section of the cortex of a young branch of *Sambucus*; (C) Annular collenchyma as seen in cross-section of the main vein of a leaf of *Nerium oleander*; (D) Lacunar collenchyma as seen in cross-section of a leaf petiole of *Petaite officinalis*; (E) Angular collenchyma as seen in longitudinal section.

a long slender cell, many times longer than wide, whereas sclereids vary in form from an approximately isodiametric to a considerably, elongated one, and some kinds of sclereids are much branched.

SCLEREIDS

Sclereids arise either through secondary sclerosis of parenchyma cells or by sclereid primorida. In leaves they may develop from epidermal cells, palisade or spongy parenchyma. In the vascular region, the sclereids develop from the derivatives of procambial or vascular cambial cells. Macrosclereids found in seed coats arise from protoderm. Randomly distributed sclereids are formed from the parenchyma cells. The development of branched and long sclereids involves remarkable intercellular adjustments. The growth of cells is called intrusive or interpositional growth and is contrasted with the coordinated growth that involves no seperation of walls. The intrusive growth is shown in detail in Fig. 5.2.

On the basis of their size, shape, nature of cell wall and mode of deposition of secondary wall material sclereids have been classified into five types.

1. Brachysclereids (Stone cells): They are shortly, roughly isodiametric or some what elongated cells. Their shape is often like that of parenchyma cells. They are found in the cortex and pith of stems, phloem and fleshy parts of fruits, e.g., Pear, Cinnamon, guava, custard apple, sapota and coconut.

2. Macrosclereids (Rod cells or Malpighian cells): They are elongated columnar sclereids. They are found in seed coats of Fabaceae, e.g., bean (*Phaseolus*), Pea (*Pisum*), soyabean (*Glycine*).

3. Osteosclereids (bone cells): They are also columnar but with swollen ends similar to a bone. They are found in seed coats (e.g., *Pisum sativum*) and some times also in the leaves of certain Dicotyledons (e.g., *Hakea*).

4. Astrosclereids: These are branched and star-shaped. They occur in the leaves (e.g., *Nymphaea*, tea) and stem cortex and leaves of *Trochodendron*.

5. Trichosclereids: These are hair-like, branched or unbranched sclereids found in the intercellular spaces of leaves and stems of some hydrophytes, aerial roots of *Monstera* and leaves of *Olea*.

Functions

1. The sclereids provide firmnes, stiffness and mechanical strength to the part of plant where they are found.
2. Sclereids also form a protective layer in the clove sclaes of garlic (*Allium sativum*).

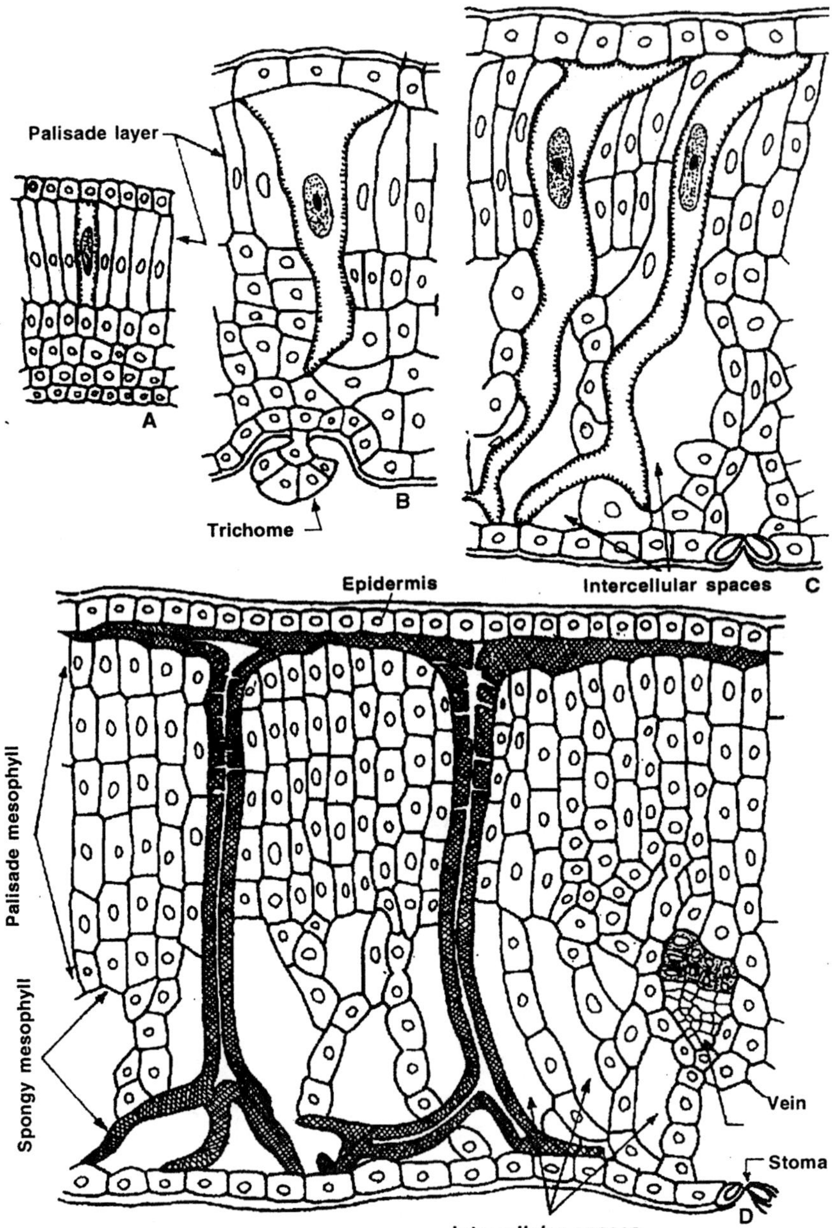

Figure 5.2: Development of sclereids in the leaf of *Osmanthus* (Oleaceae). (A–C) Differentiating sclereids indicated by large nuclei and dots along the walls; (D) Mature sclereids, indicated by cross-hatched secondary walls.

FIBRES

Fibres are much elongated elements with acute end walls, narrow lumina and thick secondary walls. In some instances they have septa and then designated as septate fibres. Some times they are so long that they reach up to 55 cm in length as in ramie (*Boehmeria nivea* — Urticaceae). Ramie fibres are longest cells in higher plants and used in cordage. Ontogenetically fibres develop from different meristems, such as the procambium, vascular cambium, ground meristem, and even from the protoderm, as in certain species of the Poaceae and Cyperaceae. Fibres may also develop from parenchyma cells, e.g., in the protophloem of many dicotyledons. The fibres formed by cambium develop from fusiform initials and elongate only little or not at all during their maturation. Fibres also show a combination of coordinated and intrusive growth.

Classification: On the basis of their position in the plant body, fibres are classified into two large groups, the xylary fibres and extra xylary fibres.

(a) **Xylary fibres:** These are also called wood fibres for the reason they are found in xylem and develop from the same meristematic tissue which gives rise to other xylem elements. On the basis of wall thickness and the nature of pits, two main types are distiguished among the xylary fibres (i). Fibre tracheids and (ii) Libriform fibres. If both occur in the same wood, the libriform fibre is longer and has thicker walls than the fibre tracheid. The pits in fibre tracheid are narrow, bordered while libriform fibres have simple pits with no borders. By and large fibre tracheids show intermediate characters when compared with tracheid and libriform fibres. Both categories of fibres may be septate. Septate fibres are distributed widely in dicotyledons and usually retain their protoplasts in mature sapwood. Septate fibres are concerned with storage of reserve food materials.

The fibres of the branch wood in the dicotyledons are characterized by the presence of inner most layer (G-layer) of the secondary wall with much α-cellulose and little of lignin. They are known as gelatinous fibres or mucilaginous fibres.

(b) **Extraxylary fibres:** These fibres are found outside the xylem such as in cortex, pericycle and phloem. When encountered in the phloem they are known as bast fibres. On the basis of their occurrence, the extraxylary fibres are categorised into (i) phloem fibres (ii) cortical fibres and (iii) perivascular fibres. Extraxylary fibres are long, spindle-like with acute or acuminate or blunt ends. Their cell wall is quite thick with simple or slightly bordered pits. At maturity, these fibres loose protoplasts and become dead. Usually these fibres remain unbranched but branched fibres may be found both in monocotyledons and dicotyledons.

The fibres are long cells, with more or less thick secondary walls and usually occur in strands. These strands constitute the "fibres" of commerce. With in a strand, the fibres overlap, a feature that imparts strength to the fiber bundles.

Economic Importance of Fibres

In commerce, the term fibre is often applied to materials that include, in botanical sense, other types of cells beside fibres and also to structures that are not fibres at all. In fact, the leaf fibres of the monocotyledons commonly include vascular elements. Cotton fibres are epidermal hairs of seeds of *Gossypium*. Similarly the vasculår system of roots of *Muhlenbeckia* and the entire plant of *Tillandsia* are used as commercial fibres. Likewise, kapok fibres of *Ceiba pentandra* originate as hairs from the inner surface of the fruit (capsule).

Commercial fibres are divided into hard fibres (with high lignin content) and soft fibres (with little or no lignin content). Hard fibres are obtained from the leaves of moncotyledons, e.g., Abaca or Manila hemp (*Musa textilis*), bowstring hemp (*Sansevieria* spp.), Sisal (*Agave sisalina*). Soft fibres (also known as bast fibres) are obtained from phloem, e.g., flax (*Linum usitatissimum*), hemp (*Cannabis sativa*), jute (*Corchorus capsularis*, *C.olitorius*), kenaf (*Hibiscus cannabinus*), sunhemp (*Crotalria juncea*), rozelle (*Hibiscus sabdariffa)*.

REFERENCES

Carlquist, S. 1988. Comparative Plant Anatomy. Holt, Reihart & Winston, New York.

Cutter, E.G. 1978. Plant Anatomy. Part I & II. Edward Arnold, London.

Eames, A.J. and L.H. MacDaniels. 1979. An Introduction to Plant Anatomy. Tata McGraw Hill, Bombay.

Esau, K. 1953. Pant Anatomy. John Wiley & Sons, New York.

Esau, K. 1977. Anatomy of seed plants. Wiley Eastern, New Delhi.

Fahn, A. 1989. Plant Anatomy. Pergamon Press, Oxford.

Jeffrey, E.C. The anatomy of Woody Plants.

Mauseth, J.D. 1988. Plant Anatomy. Benjamin Cummings, California.

Metcalfe, C.R. and L. Chalk. 1950. Antomy of Dicotyledons, 2 Vols. Clarendon Press, Oxford.

Pandey, B.P. 1993. Plant Anatomy. S. Chand & Co., New Delhi.

Rudall, Paula. 1990. Anatomy of Flowering Plants.

Steeves, T.A. and I.M. Sussex. 1989. Patterns in Plant Development (2nd edn.). Cambridge Univ. Press, Cambridge.

CHAPTER VI

COMPLEX TISSUES

The complex tissues are heterogeneous in nature, being composed of different types of cells performing diverse functions. Xylem and Phloem are complex tissues, which constitute the major part of vascular tissues. Vascular bundles form a continuous and interconnected system in different organs of the plant.

XYLEM

Xylem (Gr.: *Xylos* = Wood) is a complex tissue. It is the principal water conducting tissue in the vascular plants. It is usually spatially associated with phloem, the principal food-conducting tissue. The two tissues together are called the vascular tissue or tissues. The combination of xylem and phloem forms a continuous vascular system through out all parts of the plant including all branches of stem and root. The xylem is present in all the land plants viz., Pteridophytes, Gymnosperms and Angiosperms. Therefore, all these plants together constitute Tracheophyta.

Xylem is a complex tissue as it consists of several types of cells. The most important cells are the tracheary elements, fibres, xylem (wood) parenchyma and xylem rays. Of these only the xylem parenchyma and xylem rays are living and the rest are dead cells at maturity.

Xylem originates from two sources, (i) Procambium of the apical meristem, which gives rise to the primary xylem of the primary body, and (ii) Vascular cambium, which gives rise to the secondary xylem.

TRACHEARY ELEMENTS

The tracheids and vessel members together represent the tracheary elements. They are concerned with the conduction of water and substances dissolved in water. They are more or less elongated cells non-living at maturity. They have lignified walls with secondary thickenings and a variety of pits. The tracheids and vessel members differ from each other in that the tracheid is an imperforate cell where as the vessel member has perforations, one or more at each end, sometimes also on a side wall.

Tracheids

In tracheids, the passage of water from cell to cell occurs mainly through pit-pairs. The term, tracheid was introduced in 1863 by Sanio. A vessel, which is also termed trachea, is built of numerous vessel members that are joined one to the other by their end walls.

Cell wall: Cell wall of the tracheid is moderately thick and lignified. The secondary wall layers are deposited in different manners, which may be annular, spiral, scalariform, reticulate and pitted.

1. Annular: The lignin is deposited in the form of rings all along the walls almost at regular intervals (Fig. 6.1 G).
2. Spiral/ helical: The lignin is deposited in a manner that appears spiral shape in length. The helical thickening may be single or more than one helix may be present in a single element. The rings or helices may be arranged in a loose or dense manner (Fig. 6.1 E, F, H, I).
3. Scalariform: The lignin deposition is in the form of closely arranged transverse bands, which gives an appearance of a ladder (Fig. 6.1 J).
4. Reticulate: Wall thickening is in the form of a network (Fig. 6.1 K).
5. Pitted: The entire inner surface of the cell wall is thickened except for certain un-thickened circular areas which look-like pits (Fig. 6.1 A–D). The nature of pits may be variable. Of the above, the first formed ones are annular, spiral and reticulate thickenings. They are found in the protoxylem. Pitted elements are characteristic of the metaxylem and secondary xylem. However all these elements need not be present in the same plant. Further intermediate forms are also present.

Pitting: Transport of water and solutes takes place, in adjacent tracheids, through the pits. The pits are seen between two neighbouring tracheary elements. Between tracheary elements and fibres very few or no pit connections exist. Tracheary contact with a xylem parenchyma is through a half-bordered pit pair i.e., bordered pit on the tracheid side and a simple pit on the side of parenchyma.

On the basis of arrangement of pits of tracheary elements, three patterns are discernible.

1. Scalariform pitting: The pits are transversely elongated and arranged in longitudinal rows.
2. Opposite pitting: The pits are circular and elliptical and arranged in horizontal rows.
3. Alternate pitting: The pits are again circular or oval and arranged in oblique or diagonal rows.

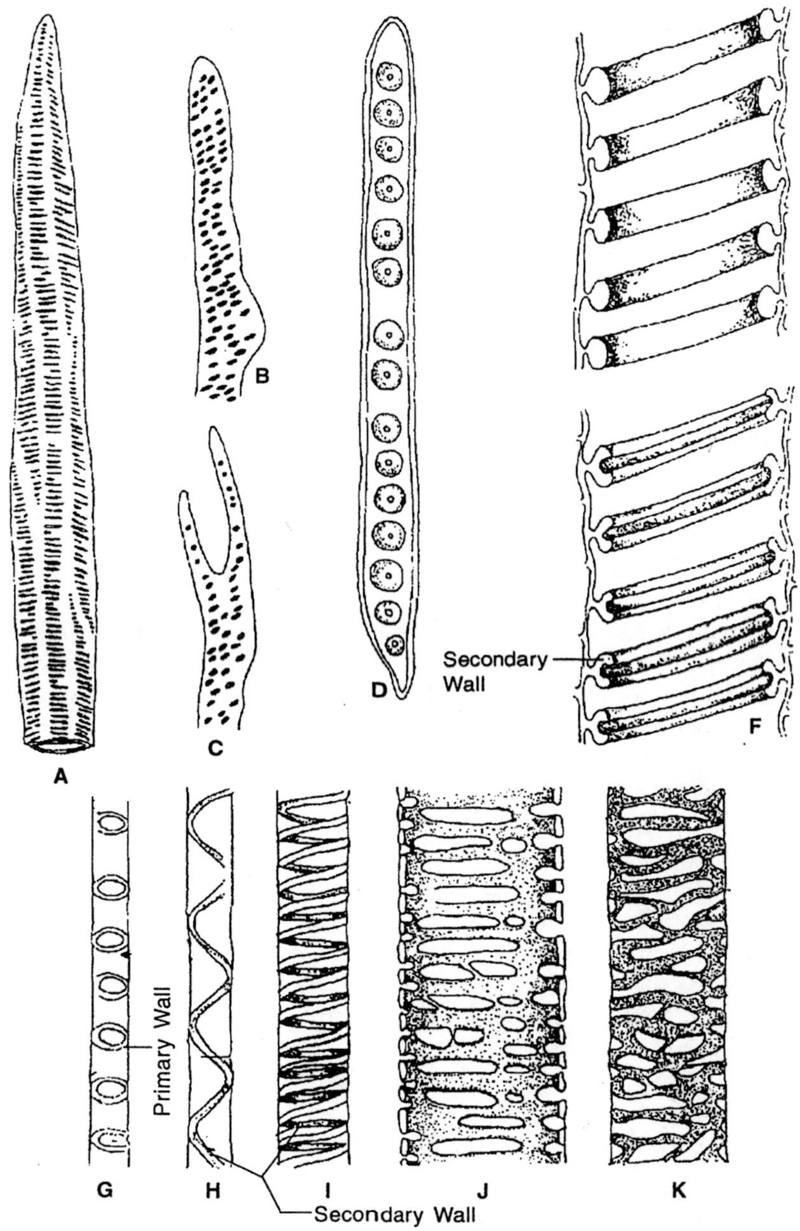

Figure 6.1: (A) Tip of tracheid of *Dryopteris*; with scalariform pitting; (B, C) Tips of tracheids of *Kingia*; (D) Tracheid of *Pinus*; (E) Portion of a longitudinally sectioned tracheary element showing the helical wall thickenings and the strips by which they are joined to the primary wall; (F) As in E but in which the helical thickening is deeply grooved; (G–K) Different types of wall thickening in tracheary elements; (G) Annular; (H) Helical; (I) Dense helical; (I) Scalariform; (J) Reticulate thickening.

Of these the scalariform pitting is the most primitive and found commonly in lower vascular plants (Pteridophytes).

Vessels or Trachea

Longitudinal series of vessel members inter connected through their perforations are called vessels. The cells are nonliving at maturity. The perforated part of a wall of a vessel member is called the perforation plate. A plate may be simple, with only one perforation, or multiperforate, with more than one perforation. The multiperforate plates are scalariform, if the perforarions are elongated and arranged parallel to each other and reticulate if the perforations form a net-like pattern.

Differentiation of Tracheary Elements

Vessels arise from procambial or vascular cambial cells in the primary and secondary xylem respectively. During the stage of cell growth and secondary wall deposition the living protoplast is concerned with the phenomena of differentiation (Fig. 6.2 A, B). The protoplast contains the full complement of organelles including a nucleus and tonoplast-bound vacuoles. The nucleus becomes polyploid and increases in size. The endoplasmic reticulum is seen in extensive profiles along and especially between the secondary wall thickenings. Microtubules are prominently displayed during the growth of cell wall. At first they are spread evenly along the entire wall, but later are concentrated in bands at the sites of secondary thickenings. After the secondary wall is deposited, the cell enters the stage characterized by the phenomena of lysis affecting the protoplast and certain parts of the cell wall. The cytoplasm is exposed to the hydrolases possibly through rupture of the tonoplast. The hydrolases also reach the cell walls and attack the primary wall parts that are not covered by lignified secondaiy wall layers. The lateral walls are partly digested (Fig. 6.2 C, D). Removal of noncellulosic components leaves a fine network of cellulose microfibrils. In the sites of the perforations, the entire primary wall part disappears.

The wall occupying the site of the future perforation is thicker than the primary wall elsewhere. The removal of this wall appears to be a gradual dissolution process indicated by the appearance of small holes before the final breakdown.

Pits: Simple and bordered pits are encountered in the secondary walls of tracheids and vessel members. The number and arrangement of these pits are highly variable, even on different wall facets of the same cell, because they depend on the type of cell contiguous with the particular wall facet. Usually numerous bordered pit-pairs occur between contiguous tracheary elements, few or no pit-pairs occur between tracheary elements and fibres;

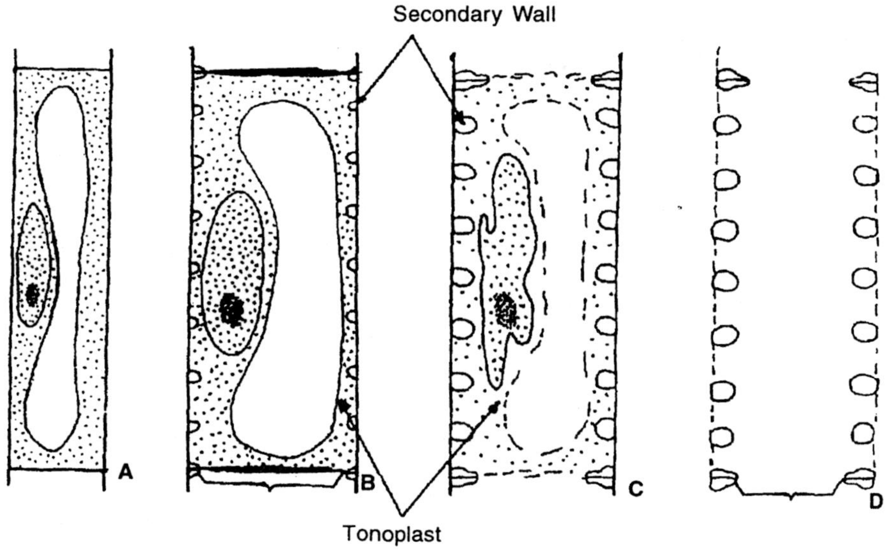

Figure 6.2: Diagrams illustrating development of a vessel member with a helical secondary thickening. (A) Cell without secondary wall; (B) Cell has attained full width, nucleus has enlarged, secondary wall has begun to be deposited primary wall at the pore site has increased in thickness; (C) Cell at stage of lysis; (D) Mature cell without protoplast.

half-bordered or simple pit-pairs are found between tracheary elements and paranchyma cells.

The pit membrane has a torus in the centre. It is surrounded by the thin part of the membrane, the margo. The torus is wider than the aperture in the border and consequently blocks the aperture when the pit-pair is aspirated. Aspirated bordered pit-pairs are common in the heart wood. The bordered pits thus act as valves preventing the entry of air into conducting tracheids, which would disrupt the water column in the latter.

Fibres: Xylem fibres are found both in primary and secondary xylem. Xylem fibres are of two types: (i) Fibre tracheids, and (ii) Libriform fibres. The fibre tracheids have bordered pits with cavities smaller than the pit cavities of tracheids or vessels in the same wood. These pits have a pit canal with a circular outer aperture and an elongated or slit like inner aperture. The pit in a libriform fibre has a slit-like aperture toward the cell lumen and a canal resembling a much flattened, funnel, but no pit cavity. Other details of xylem fibres have been given earlier.

Phylogenetic Specialization of Tracheary and Fibre Cells

The xylem holds an important position in the study of plant tissues as the structure of its elements is of extreme importance in taxonomy and phylogeny.

The specialization of treacheary elements was concomitant of the seperation of functions of conduction and strengthening in the vascular plant that occurred during the evolution of land plants. In the less specialized state, conduction and support are combined in tracheids, with increasing specialization wood evolved with conducting elements — the vessel members — more efficient in conduction than in support. In contrast, fibres evolved principally strengthening elements. Thus, from primitive tracheids two lines of specialization diverged, one toward the vessels, the other toward the fibres.

Vessels evolved independently in several taxa of vascular plants. In the primitive most Angiospermous families Winteraceae (all the seven genera — *Drimys*, *Pseudowintera*, *Bubbia*, *Belliolum*, *Exospermum*, *Zygogynum* and *Tetrathalmus*), Trochodendraceae, Tetracentraceae, Monimiaceae and Chloranthaceae have no vessels in their xylem.

Vessel members occur in the following diverse groups of plants: In the most advanced gymnosperms, the Gnetales, in dicotyledons except for the lower most taxonomic groups, in monocotyledons, in the fern *Pteridium*, in certain species of *Selaginella*, in *Equisetum* and in the root of some species of *Marsilea*. From the above it can be assumed that vessels developed independently, by parallel evolution, in each of these groups. A few Monocotyledonous families with only aquatic species are known to include plants that lack vessels completely in all their organs. This feature, however, may be a secondary one.

In some specialized dicotyledoms, such as in certain representatives of the Cactaceae, vessels in secondary xylem are replaced by so called vascular tracheids (short, imperforate cells with annular and helical thickenings. However, in such plants the reduction of vessels can not be cited as support for evolutionary reversal to primitive vesselless condition.

The evolutionary sequence of vessel member types in the secondary xylem of dicotyledons began with long scalariformly pitted tracheids similar to those still found in some lower dicotyledons. These tracheids, were succeeded by vessel members of long narrow shape with tapering ends. The cells shortened progressively, became wider, their end walls became less inclined and finally transverse. In the more primitive state the perforation plate was scalariform, with numerous bars resembling a wall with scalariformly arranged pits devoid of pit membranes. Increase in specialisation resulted in a decrease in the number of bars and finally their total elimination and the appearance of a simple perforation. The tracheids were not eliminated when vessels evolved, but underwent phylogenetic changes. They became shorter — not as short as the vessel member however — and the pitting of their walls evolved parallel to that of the associated vessel members. They generally did not increase in width.

In the specilaization of xylem fibres the emphasis on mechanical function became apparent in the increase in wall thickness and decrease in cell width. Concomitantly the pits changed from elongated to circular, the borders became reduced and eventually disappeared. The inner aperture of the pit became elongate and then slit like. Thus, the evolutionary sequence was from tracheids, through fibre — tracheids to libriform fibres.

Ecological Aspects of Xylem Evolution

According to Carlquist (1975), who has put forward an ecological approach to xylem evolution, the structural diversity in the xylem of presently living plants is a result of evolutionary changes, which are functionally adaptive to the various habitats. The adaptive changes were brought about through selective pressure in the different environments in which the various plant taxa evolved. He expects the adaptation of the anatomical features of xylem to be related to moisture availability, transpiration and to requirements for mechanical strength. Carlquist considers that great length of vessel members with scalariform perforation plates reduces resistance to water flow. Short and narrow vessel members with thick walls would increase the strength of a vessel and make it resistant to high negative pressure which occurs in xeric habitats. Scalariform perforation plates are regarded as functionally disadvantageous as compared to simple perforation plates, because of resistance of water flow. Xylem/wood with scalariform perforation is rare in plants of xeric habitats and has not been fiound in plants growing in deserts. Vines and lianas, with a requirement for efficient water conduction, possess exceptionally wide vessels, causing minimal friction to water flow. The vessel members in these plants are relatively short, apparently enhancing resistance to negative pressure. The number of vines and lianas with scalariform perforation plates is very small.

Xylem Parenchyma

Parenchyma cells are common constituents of the xylem of most plants. The parenchyma, found in secondary xylem, is classified into two types.

(a) Axial or wood parenchyma: This is derived from the fusiform cambial initials together with tracheary elements and fibres. These are some what elongated and lie in vertical series.
(b) Ray Parenchyma: These originate from ray initials of fusiform cambium and occur in radial transverse series. The cell wall of parenchyma found in primary xylem is thin and made up of cellulose but those found in secondary xylem may often have thick, lignified and pitted cell wall.

Xylem parenchyma cells store starch, oils, and many other ergastic substances of unknown function. Sometimes tannins and crystals may also be found in them.

The radial parenchyma cells are divided into two categories — procumbent and upright ray cells. A ray cell, the longest — diameter of which is oriented radially, is procumbent: the cell elongated axially is upright (Fig. 6.3 A, B). The two kinds of ray cells are often combined in the same ray, the upright cells typically appearing at the upper and lower margins of the ray. Rays composed of one kind of cell are called homocellular (Fig. 6.3 C, D), those containing procumbent and upright cells, heterocellular (Fig. 6.3 A, B).

The main function of parenchyma, both axial and ray, are storage and translocation of ergastic substances.

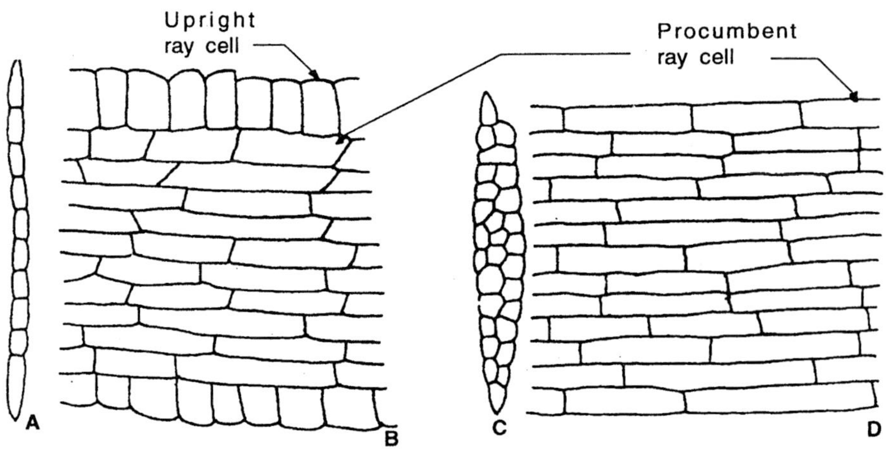

Figure 6.3: Two types of rays as seen in tangential (A, C) and radial (B, D) sections. (A, B) *Fagus grandifolia*; (C, D) *Acer saccharum*.

Protoxylem and Metaxylem

Primary xylem consists of an earlier part, the protoxylem, and the later part, the metaxylem. The protoxylem differentiates in the parts of the primary plant body that have not completed their growth and differentiation. In fact, in the shoot, the protoxylem matures among actively elongating tissues and is, therefore subjected to stresses. Its mature nonliving tracheary elements are stretched and eventually destroyed. The metaxylem is commonly initiated in the still growing primary plant body, but it matures largely after the elongation is completed. It is, therefore, less affected by the primary extension of the surrounding tissues than the protoxylem. The protoxylem usually contains only tracheary elements embedded in parenchyma that is considered to be part of the protoxylem. The metaxylem is somewhat more complex than the protoxylem and may contain fibres in addition to tracheary elements and parenchyma cells.

PHLOEM

Phloem is the principal food conducting tissue. The term phloem, was first coined by Nageli (1858) from the Greek word for bark. The phloem was also referred to as bast or leptome. Similar to the xylem, the phloem originated from two sources, (i) Procambium of the apical meristem, which gives rise to primary phloem of the primary body, and (ii) Vascular cambium, which gives rise to the secondary phloem.

The phloem is composed of sieve elements (sieve cells, sieve tubes), companion cells, phloem parenchyma and phloem fibres.

Sieve Elements

Sieve cells and sieve tube members are the two types of sieve elements. These are primarily involved with the conduction of organic solutes.

Sieve cells: They are narrow and linear with the pointed end walls. The sieve cells are joined end to end wherein their inclined end walls overlap. These are found among members of lower vascular plants and Gymnosperms (Fig. 6.4 A).

Sieve tubes: Sieve tube is a series of sieve tube members connected end to end in vertical rows, by means of sieve plates.

Sieve Areas and Sieve Plates

The most characteristic features of sieve elements are sieve areas in the walls and the disappearance of nucleus from the protoplast. The Sieve areas are wall areas with pores through which the protoplasts of vertically or laterally adjoining sieve elements are interconnected. The pores range in width from those that are comparable to plasmodesmatal pores and those several microns in diameter. In the lower vascular plants (Pteridophytes) and Gymnosperms the pores are commonly narrow and rather uniform in size in the sieve areas on different walls of the same cell (Fig. 6.4 A; 6.5 E). In the Angiosperms the size of pores varies considerably even on different walls of the same cell. Sieve areas with the larger pores usually occur on the end walls, occasionally on the side walls. The wall parts bearing the more highly differentiated sieve areas, that is, areas with comparatively large pores are called sieve plates. Each pore is lined with callose, a carbohydrate (β-1,3-glucan), that stains blue with aniline blue. As the sieve element ages, more callose accumulates. The layer with in a pore thickens and some callose also appears on the surface of the sieve area. The pore is increasingly more constricted and later completely obliterated as the sieve element becomes dormant or dies. At this stage, the callose forms a pad on the sieve area.

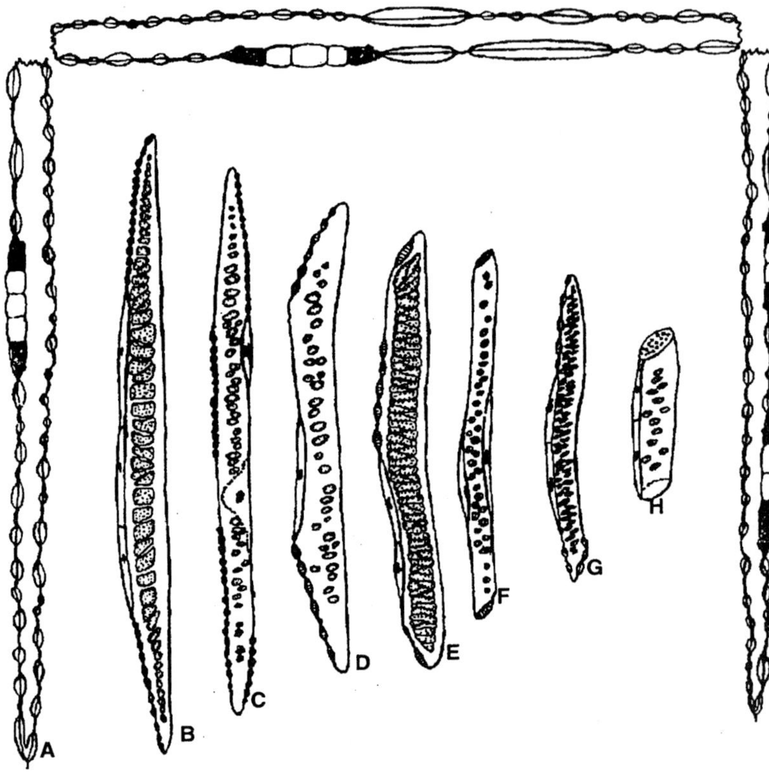

Figure 6.4: Variation in structure of sieve elements: (A) Sieve cell of *Pinus pinea* with associated rays, as seen in tangential section. Others are sieve-tube members with companion cells from tangential section of phloem of the following species; (B) *Juglans hindsii*; (C) *Pyrus malus*; (D) *Liriodendron tulipifera*; (E) *Acer pseudoplatanus*; (F) *Cryptocarya rubra*; (G) *Fraxinus americana*; (H) *Wisteria* sp. in B–G, the sieve plates appear in side views and their sieve areas are thicker than the intervening wall regions because of deposition of callose.

The callose is synthesized remarkably fast in response to injury. The deposition of callose increases during inactive season and decreases in the active season (spring time).

Sieve areas and sieve plates of Angiosperms vary widely in differentiation and arrangement (Figs. 6.4 B–H, 6.5 A–D). Long sieve tube members with much inclined end walls commonly have compound sieve plates, that is, sieve plates composed of several sieve areas. The ends of such elements are wedge-shaped and the sieve areas are borne on the oblique face of the wedge. The arrangement of the multiple sieve areas may be scalariform or less orderly.

Cell wall: Sieve element walls vary in thickness but are usually distinctly thicker than those of the surrounding parenchymatic cells. The cell walls

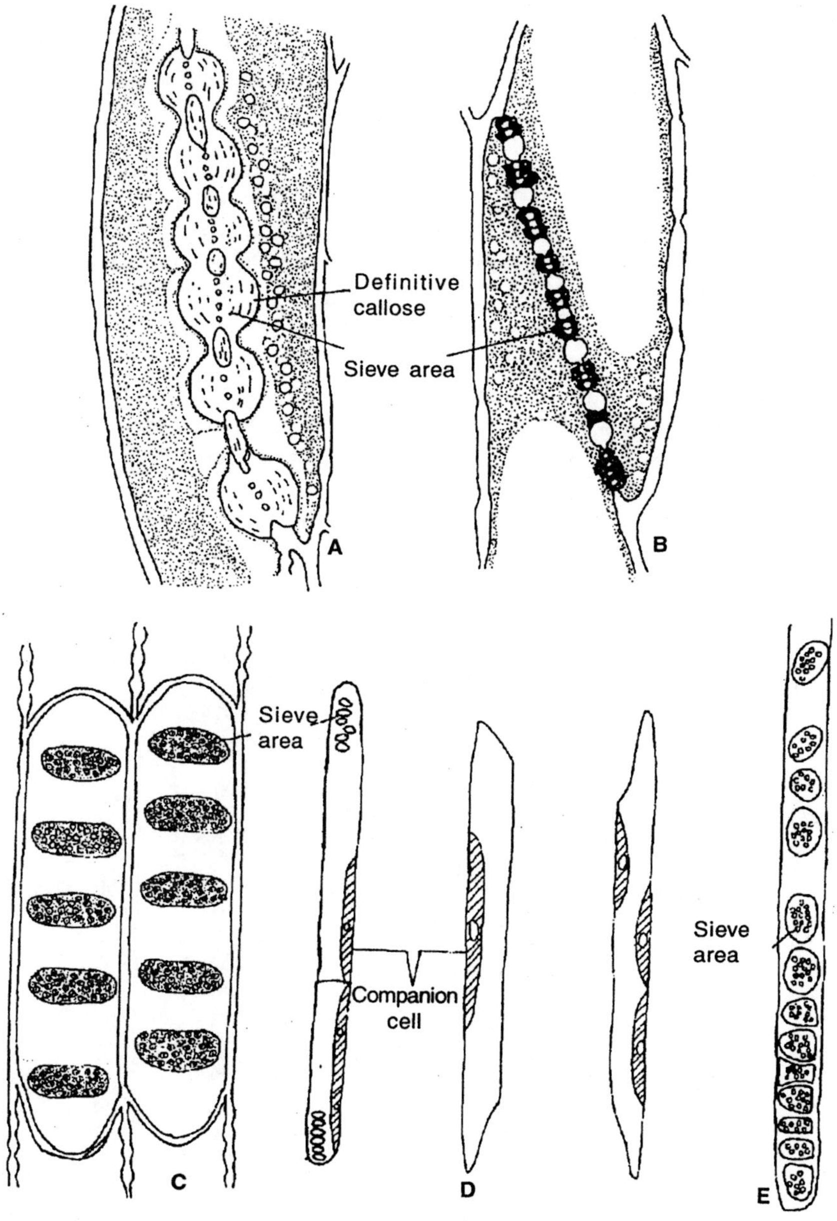

Figure 6.5: Sieve-tube members of *Vitis*. A, B. Longitudinal sections of compound sieve plates between two elements. (A) Elements of dormant state in which the plate is covered by a thick layer of callose; (B) Elements reactivated after the removal of the callus. The slime which fills the sieve areas is indicated by heavy stippling; (C) Surface view of the compound sieve plates; (D) Sieve-tube elements with companion cells; (E) Portion of a sieve cell of *Pinus*.

have a pearl luster appearance, hence they are called nacreous. The wall has interconnected microfibrils in parallel orientation. The walls of sieve elements are usually only primary and consist mainly of cellulose and protein but even the greatly thickened walls are not lignified.

Protoplast: The phloem is a unique tissue with many intriguing characteristics. One important feature of the mature sieve element is that its nucleus usually degenerates (Fig. 6.6). The degenerated nucleus may either remain as a collapsed body or disappears completely. As the maturation of sieve element takes place, the endoplasmic reticulum (ER) diminishes in amount and the ribosomes and dictysomes are not discernible. Though plasmalemma continues to function, the tonoplast breaks down and no distinction is maintained between the cytoplasm and vacuole. Only plastids and mitochondria remain in the sieve elements. The plastids occur in two basic types, one with little internal organization but accumulating starch (s-type); the other

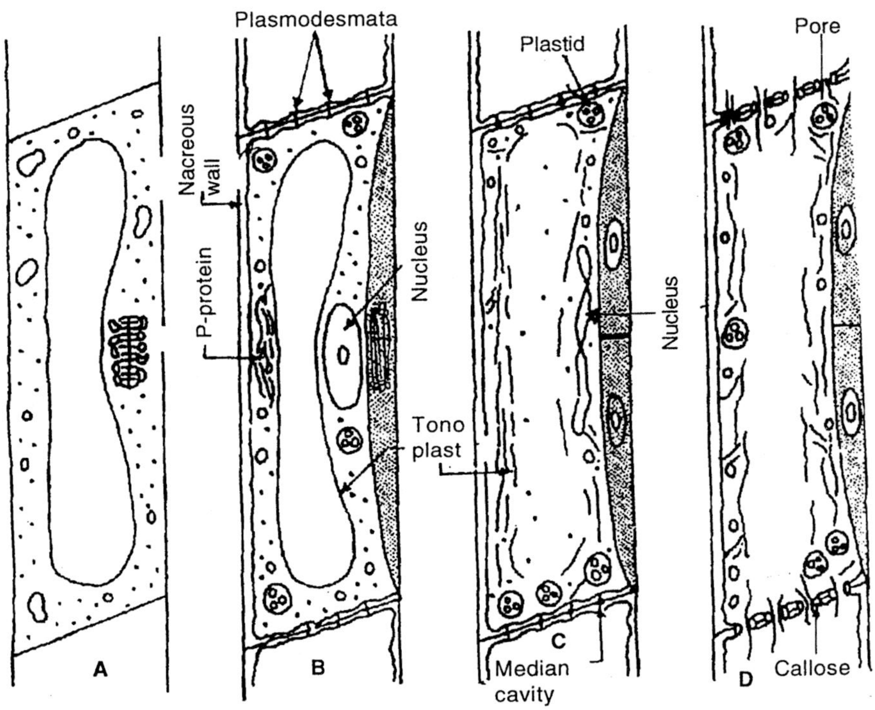

Figure 6.6: Diagrams illustrating differentiation of a sieve element. (A) Precursor of sieve element in division; (B) After division: Sieve element with nacreous wall and P-protein body; dividing companion cell precursor (stippled); (C) Nucleus degenerating, tonoplast partly broken down, P-protein dispersed; median cavities in future sieve plates; two companion cells (stippled); (D) Mature sieve element; pores in sieve plate open; they are lined with callose and contain some P-protein. No endoplasmic reticulum is shown.

containing protein as fibrils or crystals (p-type). Plastid differences in sieve elements are taxonomically useful (discussed elsewhere in this book).

Sieve elements, commonly form a proteinacous inclusion, the P-protein, formerely called slime. This protein often appears in filaments of tubular form, with subunits arranged helically. Initially the P-protein is aggregated into one or more bodies but subsequently it spreads out in the cytoplasm forming strands or networks.

Differentiation of Sieve Plate: In young sieve element, the sieve area of the incipient sieve plate is penetrated by typical plasmodesmata (Fig. 6.6 B, C), with which endoplasmic reticulum cisternae are associated. These plasomodesmata mark sites of future pores. The endoplasmic cisternae located on the two sides of a pore site are interconnected by a desmotubule extending through the plasmodesmatal canal. While the protoplast is still intact, callose appears around each plasmodesma except in the middle lamella region. The paired callose masses assume the form of platelets or cones interrupted in the centre where the plasmodesmata is located. When the nucleus degenerates plasmodesmatal canals enlarge. As the pores attain their final size the endoplasmic reticulum cisternae are removed from the pore site (Fig. 6.6 D). The plasmalemma lining the original plasmodesmatal pore remains. When the plant is cut the contents in the sieve tube readily flow out by the pressure that exists in the mature sieve element. The exudate contains some protoplast components, especially p-protein. Some of the P-protein remains lodged within the pores and forms accumulations (slime plugs) on the sieve plate.

Phylogeny of Sieve Elements

In the most primitive form the sieve elements are parenchyma cells that have undergone modifications in connection with their function. This was followed by the loss of nucleus. This protoplasmid specialization apparently resulted in the development of the interdependence of the sieve elements and parenchyma cells that retain their nucleus, i.e., the albuminous cells in Gymnosperms and the companion cells in Angiosperms. In the angiosperms the sieve elements and the companion cells develop from the same mother cell. The specialization of the sieve elements also involved the development of thick connecting strands. In Pteridophytes and Gymnosperms these strands are thin and resemble plasmodesmata, while in angiosperms they are thick and conspicuous.

The specialization of sieve elements and sieve areas has taken the following courses. (1) Gradual localization of highly specialized sieve areas to the end walls of the elements. (2) Gradual change in the position of the end wall from very oblique to horizontal. (3) Gradual change from compound sieve plates to simple ones. (4) Gradual reduction of the sieve areas on the side walls of the elements.

Sclerenchyma Cells

Fibres are common components of both primary and secondary phloem. In the primary phloem, fibres occur in the outermost part of the tissue, in the secondary, in various distributional patterns among the other phloem cells of the axial system. The fibres may be nonseptate or septate and may be living or non-living at maturity. Living fibres serve as storage cells as they do in the xylem. In many species, primary and secondary fibres are long cells with thick walls and are used as a commercial source of fibre (*Linum, Cannabis, Hibiscus, Corchorus*).

Sclereids are frequently found in the phloem. They may occur in combination with fibres or alone, and they may be present in both axial and radial systems of the secondary phloem. Sclereids typically differentiate in older parts of the phloem as a result of sclerification of parenchyma cells. This sclerification may or may not be preceded by intrusive growth of the cells. During such growth the sclereids often become branched or much elongated. The distinction between fibres and sclereids is not always sharp, especially if the sclereids are long and slender. The intermediate cell types are fibre-sclereids.

Companion Cells and Albuminous Cells

Companion cells are parenchyma cells specialized with regard to the functional association with sieve-tube members regulating the translocation. Companion cells are formed by division of sieve tube precursors so that sieve tube members and their companion cells are ontogenetically related. One or more companion cells may be associated with one sieve tube and they may occur on one or more sides of the sieve tube member. In some taxa, the companion cells appear in vertical files (companion cell strands) as a result of divisions in their precursor. Companion cells vary in size — they may be as long as the sieve tube member to which they are related or they may be shorter. These cells are strongly attached to the sieve tube members from which they usually can not be seperated even by maceration. The walls between sieve tube members and the companion cells are thin or possess many thin areas. The plasmodesmata on the companion cell side are usually branched. The length of life of the companion cells is usually the same as that of the sieve-tube member to which they are attached. The protoplasts of companion cells are those characteristic of metabolically active cells. Their nuclei and nucleoli are relatively large. The cells contain plastids, often differentiated as chloroplasts, numerous large mitochondria, ribosomes and rough endoplasmic reticulum. In most parts of the plant the sieve elements are usually considerably wider than the companion cells. However in the minor veins of the leaf the sieve elements are conspicuously smaller than their associated parenchyma cells.

These cells (companion cells or cells not ontogenetically related to the sieve elements) are called intermediate cells due to their supposed mediation in the transfer of photosynthates between mesophyll cells and the sieve elements. In the minor veins of many herbaceous dicotyledons the intermediary cells were found to develop wall ingrowths. Such cells are termed as transfer cells.

In Gymnosperms companion cells do not occur. The functional counterparts of companion cells in gymnosperms are albuminous cells (strasburgher cells). They are rarely derived from the same precursor as the sieve cell. The albuminous cells can be distinguished from the other parenchyma cells of the phloem by the presence of connections with the sieve cells and the usual absence of strarch. It was found that high respiration and acid phosphatase activities occur in the albuminous cells associated with fully differentiated sieve cells. The increased activity is restricted to those periods during which loading and unloading of sieve cells is taking place.

Primary Phloem

The primary phloem is classified into protophloem and metaphloem. The protophlem matures in plant parts that are still undergoing extension growth and its sieve elements are stretched and soon become non functional. Eventually they are completely obliterated. The metaphloem differentiates later and in plants without secondary growth, constitutes the only conducting phloem in adult plant parts.

The protophloem sieve elements of Angiosperms are usually narrow and inconspicuous, but they are enucleate and have sieve areas with callose. They may or may not have companion cells. They appear either in groups or singly among parenchyma cells or in many dicotyledones, among conspicuously elongated living cells. In many species these elongated cells are fibre primordia. While the sieve elements caese to function and are obliterated the fibre primordia increase in length, develop secondary walls and mature as fibres. Such fibres are found on the periphery of the phloem region in numerous dicotyledonous stems and are often called pericyclic fibres.

Metaphloem has more and commonly wider sieve elements than the protophloem. Companion cells are regularly present in the metaphloem of Angiosperms but fibres are usually absent. Parenchyma cells may become sclerified after the phloem caeses to conduct. The overall development and structure of phloem tissue parallel those of the xylem. But the phloem tissue is less sclerified and less persisting than the xylem tissue. Because of its usual position near the periphery of stem and root the phloem becomes much modified in relation to the increase in circumferance of the axis and is eventually cut off by the periderm. The old xylem, in contrast, remains relatively unchanged in its basic structure.

CHAPTER VII

VASCULAR CAMBIUM

The vascular cambium is the meristem that produces the secondary vascular tissues. It is a lateral meristem that develops either as longitudinal strands or as a hollow cylinder. In woody Dicotyledons and Gymnosperms the primary vascular tissues of the stem and root exist only a relatively short period, and their function is taken over by the secondary vascular tissues which are produced by the cambium. In many herbaceous Angiosperms and also in most of the recent lower vascular plants, cambium is absent or vestigial.

STRUCTURE

In certain plants, including Monocotyledons, all the cells of the procambium undergo differentiation into primary vascular tissues. In almost all the Dicotyledons and Gymnosperms a portion of the procambium remains meristematic even after the completion of primary growth and develops into the cambium of the secondary body. The cambium that arises within the bundles of the stem is called vascular cambium. The strip of fascicular cambium usually become joined by additional strips of cambium which constitute the interfascicular cambium. The interfascicular cambium is not a continuation of the procambium but develops from the interfascicular parenchyma. Hence this is called as secondary meristem from the point of view of its origin. Thus a complete cylinder of cambium is developed. In roots of most Dicotyledons and Gymnosperms the cambial cylinder develops between the primary xylem and phloem, a position that is retained through out the life of the plant.

The cambium usually consists of two types of cells.

1. Fusiform initials: These cells are very long with tapered ends. In old trunks of *Sequoia sempervirens* for example, they reach a maximum length of 8.7 mm. They are spindle-shaped.

   A fusiform initial is an approximately prismatic cell in its middle part and wedge-shaped at the ends. The tangential sides are wider than the radial. In the cambial zone, the fusiform initials and their derivatives constitute the axial system, the ray initials ray system.

One of the interesting features of cambial cells is their intense vacuolation. The walls of these cells possess primary pit-fields with plasmodesmata.

Two patterns are discernible in the arrangement of fusiform initials. When they are arranged in regular superimposed tiers, the cambium is called storied cambium.This type of cambium is found in *Tamarix, Diospyros* and *Robinia*. When the fusiform initials are irregularly aligned, the cambium is called non-storied. e.g., *Fraxinus* and *Ulmus*. This condition is apparently intermediate between the storied type and extreme non-storied type with long row of initials, such as that of *Juglans, Salix, Populus* and *Pyrus*. The Gymnosperms have cambium of non-storied type. Herbaceous plants with well-developed secondary growth for example *Solanum* (potato) have short initials that are not storied.

2. Ray initials: These cells are isodiametric. They give rise to the xylem and phloem rays oriented perpendicular to the long axis of the plant body.

Both these types of initials are larger in older trunks than in very young ones. The longitudinally oriented elements in an organ, such as the tracheary elements, fibres, xylem and phloem parenchyma and the sieve elements, develop from the fusiform initials. The cells of the vascular rays which are oriented horizontally in the organ, develop from the ray initials.

Types of Cambium

Two types of cambium can be distinguished on the basis of the arrangement of the fusiform cells as seen in tangential section.

1. Storied or stratified cambium (Fig. 7.1 A) in which the fusiform initials are arranged in horizontal rows so that their ends are approximately at the same level. The length of these initials is found to vary between 140 μ and 520 μ. This type of cambium is found in *Tamarix* and *Robinia* for example.
2. Non-storied or non-stratified cambium (Fig. 7.1 B) in which the fusiform initials partially overlap one another. Initials of this type of cambium is found to vary between 320 and 2300 μ in many dicotyledons. In vesselless dicotyledons the fusiform initials may reach a maximum length of 6200 μ. In the gymnosperms initials of 1000–8700 μ long were reported.

Cambial initals divide periclinally and produce secondary xylem and phloem. At one time a derivative cell is produced toward the xylem, at another time toward the phloem, although not necessarily in alternation.

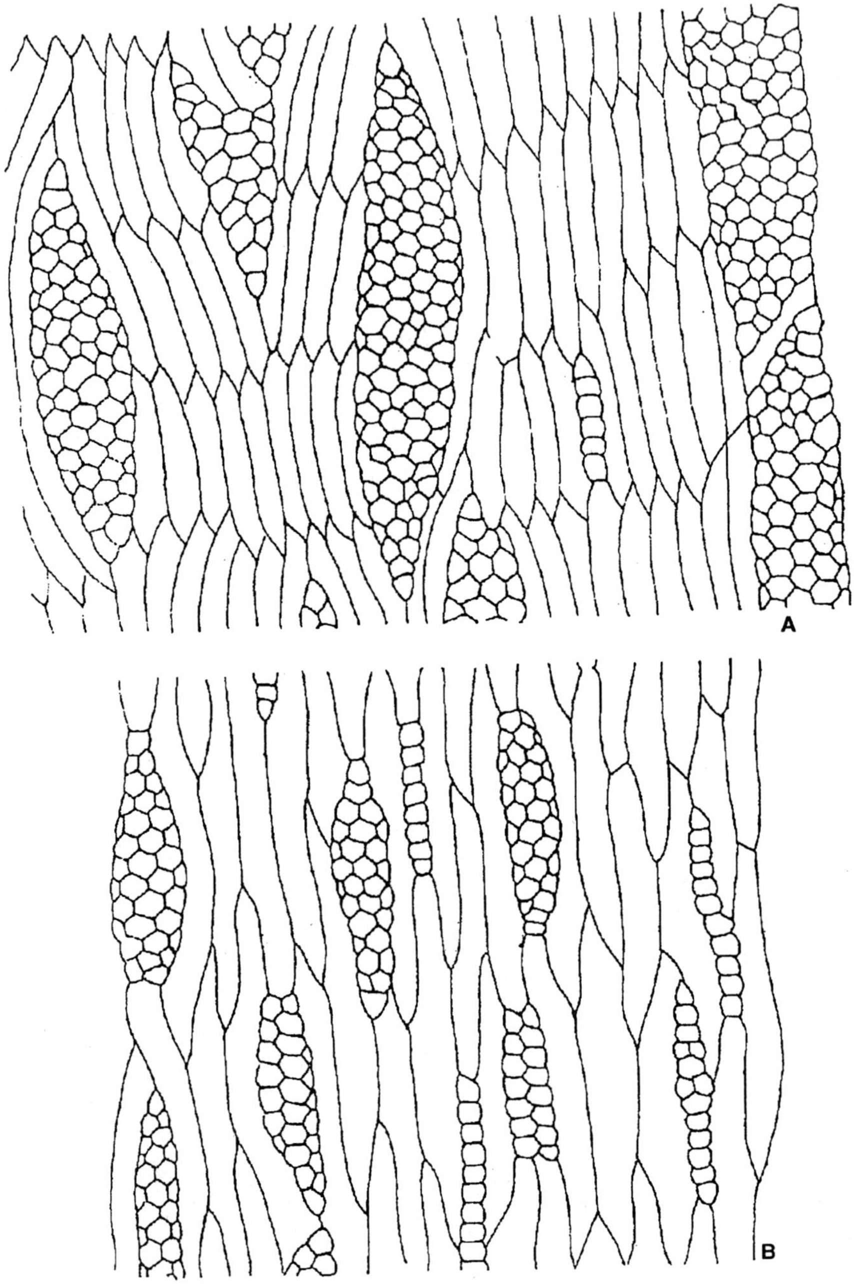

Figure 7.1: Tangential sections of different cambial types. (A) Storied cambium of *Robinia*; (B) Non-storied cambium of *Fraxinus*.

Cambial divisions that add cells to the secondary vascular tissues are called additive divisions.

As a result of the secondary thickening the circumference of the xylem cylinder increases. Together with this the cambium also increases in circumference by division of cells. The divisions increasing the number of initials are called multiplicative divisions. In cambia with short fusiform initials, the multiplicative divisions are mostly radial anticlinal (Fig. 7.2 A). Thus two cells appear side by side where one was present formerly and each enlarges tangentially. In herbaceous and shruby Dicotyledons the anticlinal divisions are frequently lateral, that is, they intersect twice the same mother cell wall (Fig. 7.2 B). Long fusiform initials divide by more or less inclined anticlinal walls (pseudotransverse division) (Fig. 7.2 C–E) and each new cell elongates by apical intrusive growth (Fig. 7.2 F, G). As a result of this growth the new sister cells come to lie side by side in the tangential plane and they thus increase the circumference of the cambium. Increase in the number of vascular rays is also an important factor contributing to the increase of the cambium cylinder in circumference. New ray initials are formed from fusiform initials by transverse division of an entire cambium cell or sometimes of part of the cell only.

The multiplicative and additive divisions commonly occur toward the end of the maximal growth concerned with the seasonal production of xylem and phloem. In plants with non-storied cambia this timing in divisions means that the cambium contains, on the average, shorter fusiform initials at the end of the season than earlier. Subsequently the new cells elongate so that the average length of the initials increases until a new period of divisions ensues at the end of the growth season.

The first additions from the initials layer may be made toward either the xylem or the phloem, the variation depending on plant species. Generally, however, more cells are added to the xylem than the phloem. Cell production toward the xylem is about four times to that toward the phloem. A greater difference was observed in a conifer. In a vigorously growing *Thuja occidentalis*, 12 to 16 new cells were formed on the phloem side and 100 or more on the xylem side.

Cambial Activity

The pattern of radial growth is correlated to the rate of cambial activity. The variation in the number of cells across the cambial zone indicates the balance between the rate of cell division and the rate of differentiation of the new derivatives. During the active period of cambium the process of division is faster than the differentiation. As a result cambial zone becomes wide. With the beginning of differentiation, a balance is established and the width of the zone remains more or less constant. When the rate of division becomes

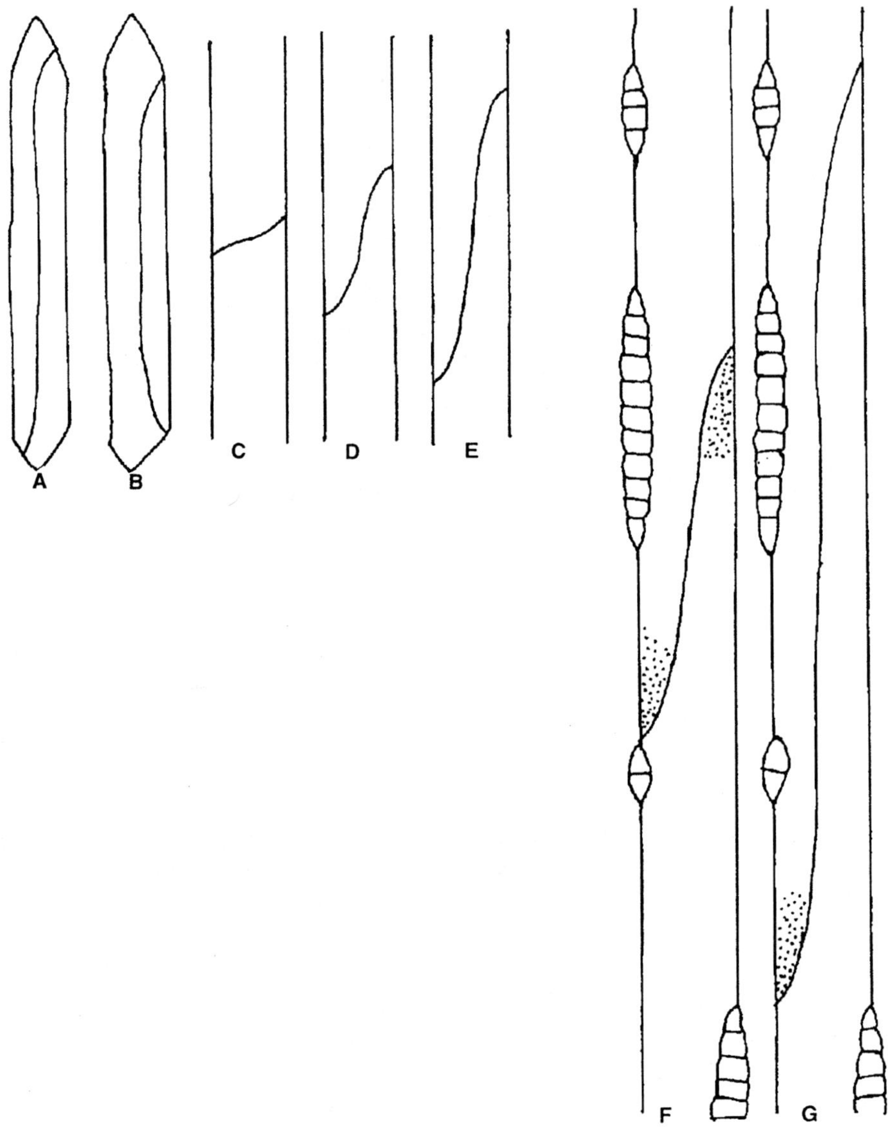

Figure 7.2: Division and growth of fusiform initials. Initial divides. (A) by radial anticlinal wall; (B) by lateral anticlinal wall; (C–E) by various anticlinal divisions; (F, G) oblique anticlinal division is followed by apical intrusive growth (growing apices are stippled).

reduced and the differentiation proceeds faster than division, the cambial zone becomes narrow.

The vascular cambium shows great variation in the period and intensity of activity. These variations are the result of many internal and external factors. There are plants whose cambium is active through out the entire life of the plant. The cambial cells divide continuously and the resulting cells undergo gradual differentiation to form the xylem and phloem elements. This type of activity is usually found in plants growing in tropical regions. In regions with definite seasonal climates the cambium caeses its activity with the onset of unfavourable conditions, usually the autumn. It enters into a dormant state which may last from the end of summer till the following spring. In spring the cambium again becomes active. From an anatomical point of view the commencement of the cambial activity usually consists of two stages: (1) the cambial cells expand radially, and (2) the cells begin to divide.

Various methods are used to determine the cambial activity: (1) The ease by which the bark can be peeled is often used as an indication of cambial activity, since with the enlargement of the cambial cells their radial walls usually become weakened, so that the bark of the stem and the roots can easily be peeled. The seperation usually occurs in the region of young xylem cells which have already reached their maximum diameter, but which still have their primary walls. (2) Examination of a cross section cut in the region of the cambium and the tissue adjacent to it, is the second method. The number of cell layers in the cambial zone and of the not yet fully differentiated xylem elements is taken as an indication of the rate of cambial activity, in this method. (3) The rates of both cell division and differentiation can be precisely determined in this method. It involves the application of radio active carbon to photosynthesizing plants. When the cambium is active the radio active carbon is incorporated in the newly formed cell walls and it can easily be detected by autoradiographic techniques.

Within the limits set by the genotype, cambial activity is dependent on factors like growth substances, temperature, photoperiod and soil moisture.

The activity of the cambium usually starts below the sprouting buds and from there it spreads downwards. In deciduous porous dicotyledons it was found to be slow. Divisions in the cambium may not begin at the base of the trunk until several weeks after they begin in the twigs. In ring porous dicotyledons, the downward spread of cambial activity is very rapid. In conifers the condition is intermediate and the spread of divisions in the cambium takes about a week.

A certain level and combination of growth regulating substances form the stimulus for the reactivation of the cambium. However in most plants a certain period of dormancy has to be completed before the cambium can be reactivated by known treatments like the application of growth regulators and increased day length. Two main groups of regulators, gibberellins and auxins, have been shown to effect the cambial activity. Gibberellins induce rapid cell division which are not followed by differentiation in the cambium. Auxins cause rapid cell differentiation. The simultaneous effect of both these groups of substances seems to result in the appearance of normal cambial activity.

Wound Healing

One of the important functions of the cambium is to form callus or wound tissue over wounds. This tissue consists of masses of soft parenchyma tissue which are rapidly formed on or below the damaged surface of stems or roots. Callus may be formed by the division of parenchyma cells of the phloem, cortex or vascular rays, but it is mostly formed by the cambium. The outer cells of the parenchymatous masses become suberized or a periderm develops in them. Below this protective layer a reorganized cambium produces new vascular tissues.

Callus is also developed, at the beginning of the growth season, on the circumference of wounds caused by pruning. With the continued production, by the cambium of secondary xylem in the undamaged area around the scar and because of the eventual fusion of the cambial layers, the wound becomes completely covered. The production of the secondary xylem continues so that the layer covering the wound is continually thickened.

CHAPTER VIII

LATICIFERS

Laticifers are cells or series of connected cells that contain latex. The laticifers may be simple or compound in origin. The simple laticifers are single cells, the compound laticifers are derived from a series of cells. Laticiferous cells are parenchymatous cells. The primary walls are thin, devoid of pits and made up of cellulose and large proportion of hemicellulose with pectic substances.

Latex is suspension or in certain cases an emulsion, of many small particles in a liquid with a varying refractive index. The chemical composition of the suspended or emulsified particles of the latex differs in different species. Among the suspended material rubber particles $(C_5H_8)_n$, waxes, resins, proteins, essential oils, mucilages and in *Euphorbia* species, variously shaped strach grains can be found. Latex, like cell sap, also contains salts, organic acids, and other substances in true solution, certain plants contain in the latex sugars (the Asteraceae), tannins (*Musa*), alkaloids morphine, codeine and papaverine (*Papaver somniferum*), and in *Carica papaya*, a proteolytic enzyme, papain, is present. The colour of the latex may be white and milky (*Euphorbia, Lactuca, Asclepias, Calotropis*), yellow-brown (*Cannabis*), yellow (*Argemone*), yellow-orange (*Papaver*) or colourless (*Morus*).

The function of latex in the plant is not clear. Haberlandt (1918) believed that latex to be of nutritional value in many plants. Sperlich (1939) regarded latex as a reserve material. The most acceptable theory is that the latex contains metabolic by-products and the laticifers represent a secretory system.

Laticifers occur in various tissues and organs of the plant body but may be restricted to phloem. Laticifers are present in about 20 angiospermous families, e.g., Euphorbiaceae, Moraceae, Asclepiadaceae, Apocynaceae, Papaveraceae etc.

TYPES OF LATICIFERS

According to de Bary (1977) laticifers are divided into two types: (1) Non-articulated laticifers also known as simple laticifers or latex cells, (2) Articulated laticifers also known as latex vessels or compound laticifers.

1. Non-articulated laticifers: They are single cells. They arise from some of the meristematic cells of the embryonic stage or from meristem of the developing shoot. As the plant grows, the latex cells also develop and enlarge. They enter other tissues. The latex cells show apical growth and branch out (Fig. 8.1 A). Non-articulated latex cells show multinucleate (coenocytic) condition with scanty cytoplasm and thicker walls than the neighbouring cells. They are found in the families Apocynaceae, Asclepiadaceae, Euphorbiaceae, Moraceae and Urticaceae. The non-articulated laticifers may be branched (e.g., *Euphorbia, Nerium, Ficus, Asclepias*), or unbranched (e.g., *Vinca, Urtica* and *Cannabis).*
2. Articulated laticifers: They are compound in origin i.e., they develop from a series of cells whose cross walls break down completely or partially (Fig. 8.2 B). The articulated laticifers are also coenocytic. They also arise from the embryonic stage or from meristem of developing shoot. Articulated laticifers are found in the families Compositae, Convolvulaceae, Papaveraceae, Euphorbiaceae, Caricaceae, Sapotaceae, Liliaceae and Musaceae.

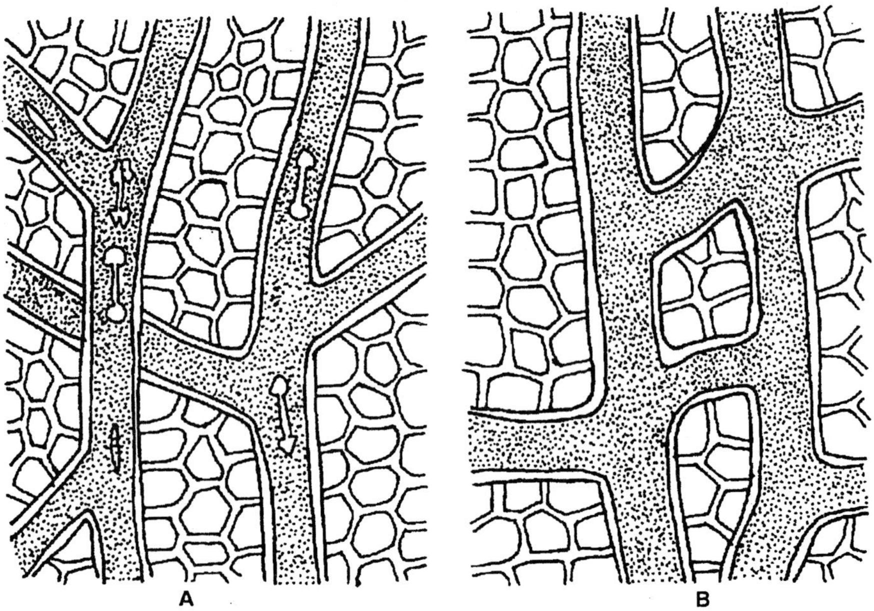

Figure 8.1: Laticifers. (A) Non-articulated; (B) Articulated.

ECONOMIC IMPORTANCE

Commercially, latex is the most useful one among the plant secretions. Rubber is obtained from the latex of bark of *Hevea brasiliensis* (para rubber plant) of Euphorbiaceae. *Manihot glaziovii*, *Ficus elastica* (Indian rubber plant), *Cryptostegia grandiflora* and *Parthenium argentatum* (Guyayule rubber) also give rubber but their yields are very low when compared to *Hevea*. The latex of *Papaver somniferum* (Opium poppy) contains alkaloids morphine, codeine and papaverine which are used in Anasthesia. Guttapercha is obtained from the latex of *Palaquim gutta* which is employed in the manufacture of artificial dentures, golf balls, underground and underwater cables. Chickle obtained from the latex of *Manilkhara achras* (= *Achras sapota*), is used in the preparation of chewing gum. Latex secreted by the plants of the Euphorbiaceae contains hydrocarbons and other substances similar to Petroleum.

CHAPTER IX

EPIDERMIS

Epidermis (Gr.: *Epi* = upon; *derma* = skin) represents the protective covering layer of primary body of land plants. Epidermis is normally one layered (uniseriate). However, in some plants the protodermal cells divide periclinally in certain cases to form multiseriate (multiple) epidermis, e.g., *Nerium*, *Ficus*, *Peperomia*. In *Ficus* leaves, multiple epidermis is found on adaxial (upper) surface only but in *Nerium* it is found on both surfaces. In *Peperomia*, the epidermis consists of 14 to 15 layers. The velamen, the special absorbing tissue of Orchid aerial roots, is also a multiseriate epidermis.

Origin: The outer most layer of the apical meristimatic cells, called protoderm or dermatogen divides anticlinally and by further division gives rise to uniseriate layer. In plants showing Tunica — Corpus organization of the shoot apex, the epidermis is derived from the tunica.

Epidermis is distinguished into three components: (1) Epidermal cell complex; (2) Stomatal complex; (3) Trichome complex.

1. Epidermal cell complex: It forms bulk of epidermis. It consists of (a) Epidermal cell, (b) Coastal and (c) Special epidermal cells.

(a) Epidermal cell: The ordinary cells of the epidermis vary in shape, size and arrangement, but they are always closely attached to form a compact layer devoid of intercellular spaces. The cells are tabular in cross-section and isodiametric or linear in surface view. In surface view, the walls of epidermal cells are straight, curved, wavy or deeply sinuated (u, v or w-shaped).

Epidermis consists of living cells. Cells have a large central vacuole and thin peripheral uninucleate cytoplasm, which contains chromoplasts and leucoplasts and occasionally chloroplasts. The chloroplasts are found in guard cells of stomata of aquatic plants and plants growing in humid climates. The anthocyanin rich chromoplast is found in vacuolar sap of petals and tender coloured leaves.

Epidermal cell walls vary in thickness in different plants and in different parts of the same plant. In the thin — walled epidermis, the outer wall is frequently thickest.

Cutin, a waxy substance, is usually present in the outer walls of epidermal walls. This substance is found within the cell wall, i.e., interfibrillar and intermicellar spaces of the cellulose (cutinisation), and also constitutes a special layer — the cuticle — on the outer surface of the cell wall (cuticularization). The cuticle is usually absent from the actively growing parts of the roots. The thickness of cuticle varies in different plants. Plants in dry habitats show thicker cuticles. The surface of the cuticle may be smooth, rough, ridged or furrowed. The cuticle on the surface of the epidermis often penetrates, to a certain extent, between the radial walls of the epidermal cells. The cuticle is impervious to water and restricts transpiration besides providing protection. It is also resistant to microorganisms as well as decay.

Wax occurs as surface deposit in the form of 'bloom' in many glaucous leaves and fruits, e.g., *Saccharum* stem, lotus leaf, and grapes. The wax may be deposited in large or small flakes or granules, rods, sheets etc. Knowledge about the wax coating of plant parts is vital in the use of herbicides and fungicides, as waxes resist wetting by sprays.

In some cases, lignified epidermal cells have also been observed as in case of *Cycas* leaves and *Pinus* needles.

(b) Coastal cells: These cells are confined to areas overlying the veins and differ from other epidermal cells. They are in general, elongated in shape and oriented parallel to the long axis of the veins concerned.

(c) Special epidermal cells:

1. Lithocysts: The epidermal cells of plants belonging to families like Urticaceae, Cucurbitaceae, Moraceae and Acanthaceae contain deposition of calcium carbonate crystals, arranged in the form of bunches, like those of grapes. Such depositions are called cystoliths and cells containing them are called lithocysts. Lithocysts are usually larger than other epidermal cells (Fig. 9.1).
2. Myrosin cells: Certain members of family Brassicaceae contain special sac — like cells scattered in the epidermis. These cells are filled with an enzyme, myrosin, and are called myrosin cells.
3. Sclerotic cells: The epidermis of seed coats of *Pisum* and *Phaseolus* of family Fabaceae and the clove scales of *Allium sativum* of family Liliaceae is made up of layer of sclereids.

In Poaceae (grasses) small cells (silica cells) filled with silica solidified into bodies of various shapes and cells with suberized walls (cork cells) occur in pairs which alternate with elongated epidermal cells (Fig. 9.3 B).

Bulliform cells: Some epidermal cells found in groups in the leaves of certain monocots, especially grasses (maize, sorghum, sugarcane etc.) are comparatively larger, thin walled and highly vacuolated. These are called Bulliform (meaning bubble like) cells. These are commonly found in the

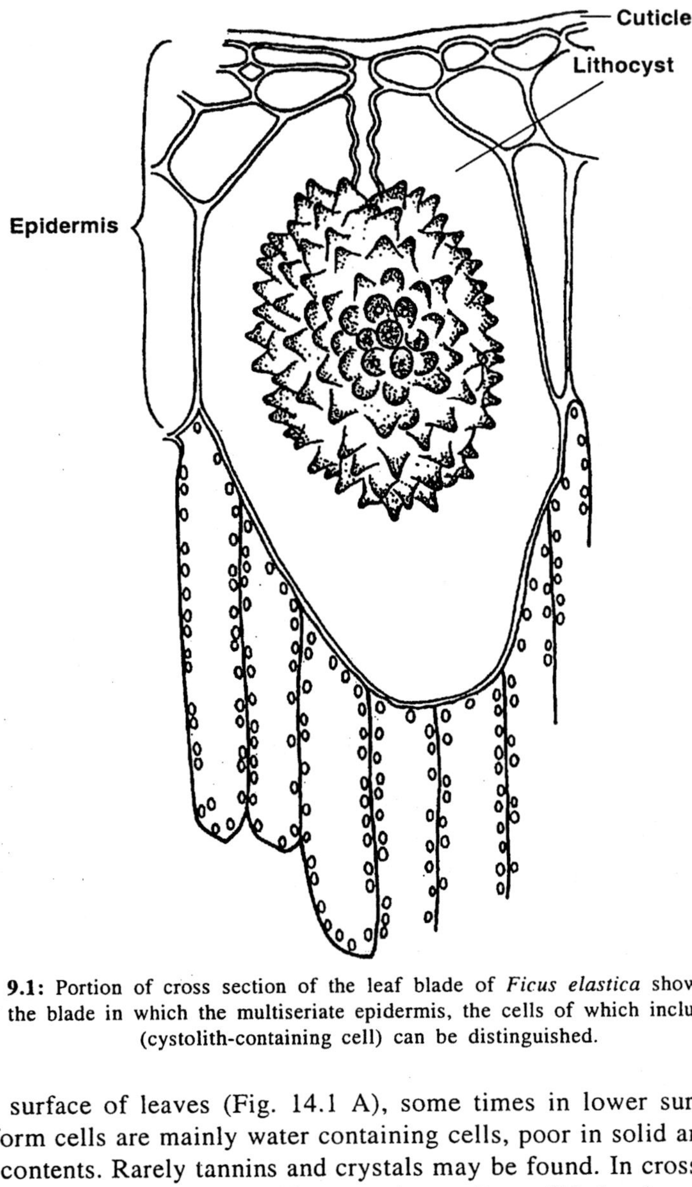

Figure 9.1: Portion of cross section of the leaf blade of *Ficus elastica* showing adaxial side of the blade in which the multiseriate epidermis, the cells of which include lithocyst (cystolith-containing cell) can be distinguished.

upper surface of leaves (Fig. 14.1 A), some times in lower surface also. Bulliform cells are mainly water containing cells, poor in solid and chlorophyll contents. Rarely tannins and crystals may be found. In cross sections, these appear as a fan-like band because the median cell being largest in size. There are three views regarding their function. (a) These are concerned with the unrolling of developing leaves by undergoing sudden and rapid expansion bringing about unrolling of lamina. (b) These are supposed to play an important role in hygroscopic opening and closing movements of mature leaves, due to changes in turgor. They are also called motor cells by some scientists. (c) These are concerned only with water storage and have no other functions.

STOMATAL COMPLEX

The epidermis shows numerous openings or pores called stomata or stomates (Gr. *Stoma* = mouth). The stomata consist of two guard cells enclosing a pore between them. The guard cells are kidney-shaped in Dicotyledones but dumb-bell-shaped in the members of Poaceae. Their wall is unevenly thickened. The wall surrounding the pore is thickened and inelastic due to presence of secondary layer of cellulose but remaining part of the wall is thin, elastic and permeable. Each guard cell has cytoplasmic lining and a large central vacuole filled with cell sap. The cytoplasm contains a nucleus and a number of chloroplasts, often poorly developed but capable of photosynthesis. Electron microscopic studies have revealed that this chloroplast has fewer and less organised lamellae than those of mesophyll cells. The guard cells are also covered with cuticle which extends to the inner wall forming the boundary of the pore and the substomatal chamber. mitochondria, dictyosomes, endoplasmic reticulum, and ribosomes are found in guard cells. Epidermal cells abutting guard cells are called subsidiaries (subsidiary or accessory cells). The stoma with its subsidiaries is termed stomatal complex or stomatal apparatus. A cavity is present beneath the stoma, known as substomatal chamber which communicates with the intercellular spaces of the mesophyll.

Van Cotthem (1970) divided the stomata into 8 types. They are:

Anomocytic (Ranunculaceous type) (Fig. 9.2 A): Stoma surrounded by a limited number of cells that are indistinguishable in size, shape or form from those of remainder of epidermis, e.g., Ranunculaceae, Malvaceae, Papaveraceae, *Tridax procumbens*, *Boerhavia diffusa*, *Cleome* sp.

Anisocytic (Cruciferous type) (Fig. 9.2 B): Stoma surrounded by three cells of which one is distinctly smaller than the other two, e.g., Brassicaceae, Solanaceae.

Paracytic (Rubiaceous type) (Fig. 9.2 C): Stoma accompanied on either side by one or more subsidiary cells parallel to the long axis of the pore and guard cells, e.g., Rubiaceae, Magnoliaceae.

Diacytic (Caryophyllaceous type) (Fig. 9.2 D): Stoma enclosed by a pair of subsiadiary cells whose common wall is at right angles to the guard cells, e.g., Caryophyllaceae, Acanthaceae.

Tetracytic (Fig. 9.3 A): Four subsiadiary cells are present, two lateral and two terminal, e.g., most of the families of Monocotyledones.

Actinocytic: Stoma surrounded by a circle of radially elongated subsidiary cells, e.g., Commelinaceae, Musaceae.

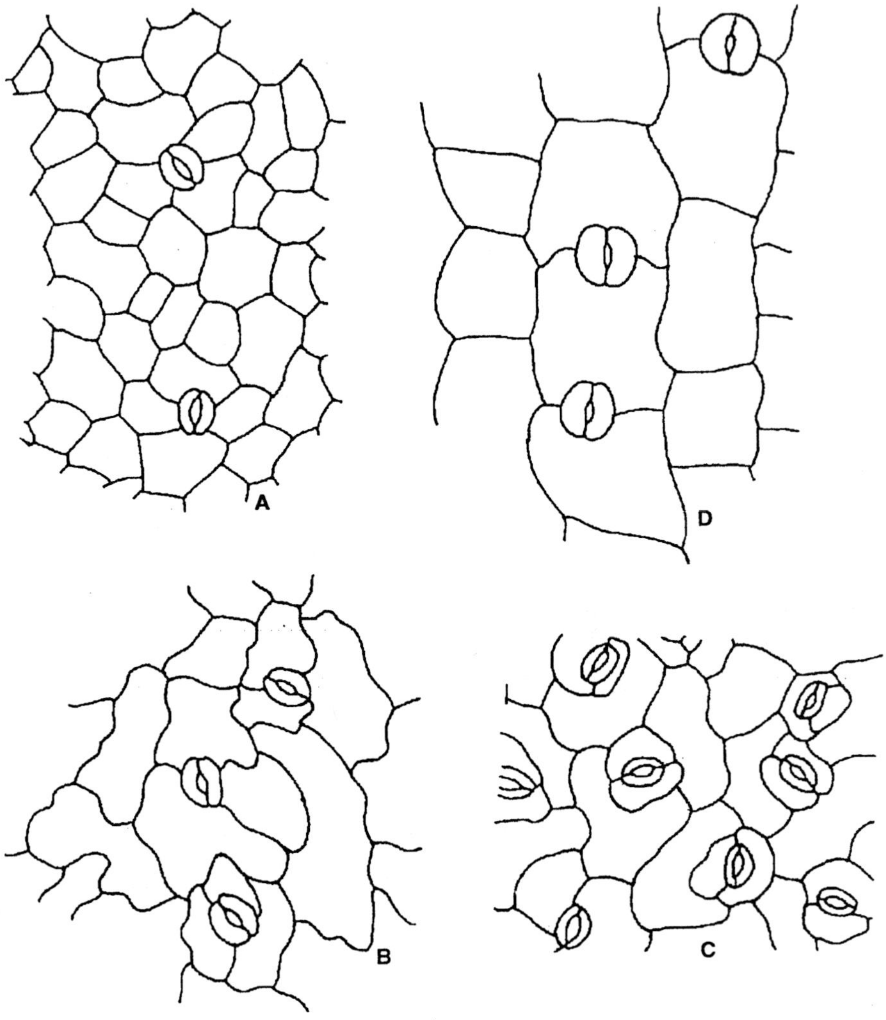

Figure 9.2: Epidermis in surface views illustrating patterns formed by guard cells and neighbouring cells. (A) *Pelargonium* — Anomocytic; (B) *Brassica* — Anisocytic; (C) *Dianthus* — Diacytic; (D) *Vigna* — Paracytic.

Cyclocytic: Stoma surrounded by four or more subsidiary cells which form a ring around each stomata, e.g., Tribe Laguncularieae of the family Combretaceae.

Hexacytic: Stoma accompanied by six subsidiary cells consisting of two lateral pairs parallel to the long axis of the pore and two polar (terminal cells), the second lateral pair as long as the stomatal complex.

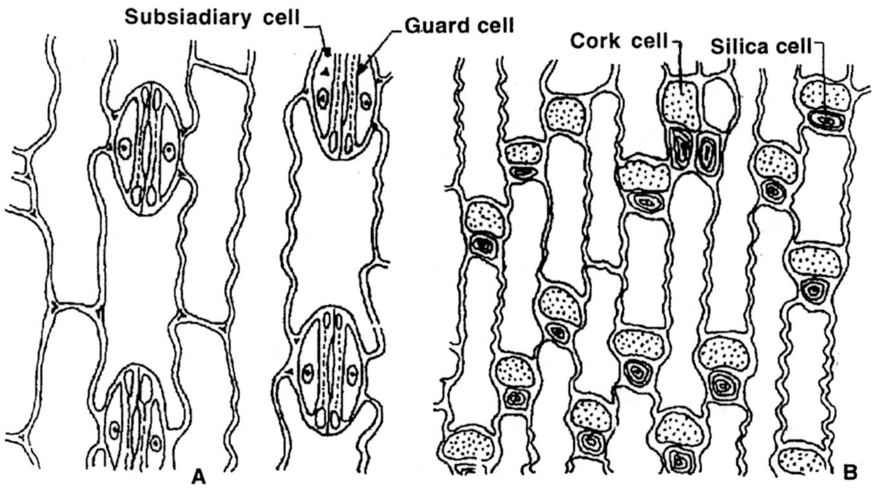

Figure 9.3: Epidermis of *Saccharum* in surface view. (A) Lower epidermis of leaf with stomata; (B) Epidermis of stem with cork cells and silica cells.

Distribution of Stomata: Stomata are present on all parts of the shoot including the floral parts. They may be non-functional on some parts of the plant body. However, they show their maximum concentration in the green leaves. Stomata are absent in the submerged plants. On the basis of stomatal distribution, the following three categories of leaves are recognisable: (a) Amphistomatic: Stomata present on both surfaces of leaf, e.g., most of the herbs, (b) Hypostomatic: Stomata confined to the abaxial (lower) surface of leaf, e.g., most of the trees, (c) Epistomatic: Stomata confined to the adaxial (upper) surface of leaf, e.g., *Nymphaea*.

Functions: Stomata play a major role in photosynthesis, respiration and transpiration. In xerophytic .plants like *Nerium* stomata are confined to stomatal crypts and covered by hairs.

TRICHOMES

Epidermal outgrowth is referred to as trichome (Gr.: a growth of hair). Trichomes vary in size and complexity and include hairs, scales and other structures. They may be glandular or non glandular. Trichomes are found both on vegetative as well as reproductive parts and may be living or dead, ephemeral or persistent. Foliar trichomes are important in regulation of moisture exchange with the atmosphere, covering hairs for plant defence against phytophagous insects, secretory trichomes, for providing chemical defence againsts animals and stinging hairs for protecting plants from animals. A study on vesiculated hairs on leaves of *Atriplex halimus* and *A. portulacoides* suggests that this type of hair may serve to remove salts from the leaf tissue and thus prevent an accumulation of toxic salts in the plant (Fig. 9.4 E). Uphof (1952) classified trichomes on the basis of morphological characters. The important categories are:

Non-glandular Trichomes

These are of two types.

Unicellular trichomes: These are simple epidermal elongations which may be straight or curved, bifurcate at apex, y-shaped (e.g., *Buddleja*) or T-shaped (e.g., *Lobularia, Dicranus*). Unicellular trichomes of *Cannabis sativa* are highly silicified with warty protuberances.

Multicellular trichomes: They may be branched (Fig. 9.4 A) or unbranched consisting of a single row of cells (uniseriate) or several rows of cells (multiseriate). The uniseriate trichomes may be vesiculate type (terminal cell bladder like as in *Chenopodium*) or capitate or sub-capitate type. The hairs may be star-shaped (stellate, e.g., *Althaea, Styrax* Fig. 9.4 B, C) or tree-like (dendroid, e.g., *Mimosa* and *Platanus*). Some times these hairs consist of disc-like plate of cells arranged on a short stalk such as peltate hairs or scales of *Olea* (Fig: 9.5 A, B). The multicellular shaggy hairs of *Portulaca* consists of several contiguous row of cells (Fig. 9.4 D).

Glandular Trichomes

The glandular trichomes include uni or multicellular trichomes. Glandular trichomes are involved in the secretion of various substances, e.g., salt solution, sugar solutions (nectar), terpenes and gums (polysaccharides). The secreting trichomes are often called glands.

(a) Salt-secreting trichomes. (i) Bladder-like hairs consisting of a large secretory cell on top of a narrow stalk consisting of one or sometimes a few cells and a basal cell as seen in *Atriplex.* (ii) Multicellular glands consisting of several secretory cells and basal collecting cells. It includes chalk glands as in *Plumbago capensis* (Fig. 9.5 H)and salt glands in *Limonium* (Fig. 9.6 E), *Avicennia* and *Tamarix* (Figs. 9.5 C, 9.6 E). In these glands the cytoplasm is dense, rich in mitochondria, ER and Golgi bodies and has many vesicular structures. The salt solution is actively secreted onto the surface of the secretory cells. Trichomes that secrete and aqueous solution containing some inorganic and organic substances were termed trichome — hydathode, e.g., *Cicer arietinum* (Fig. 9.6 B).

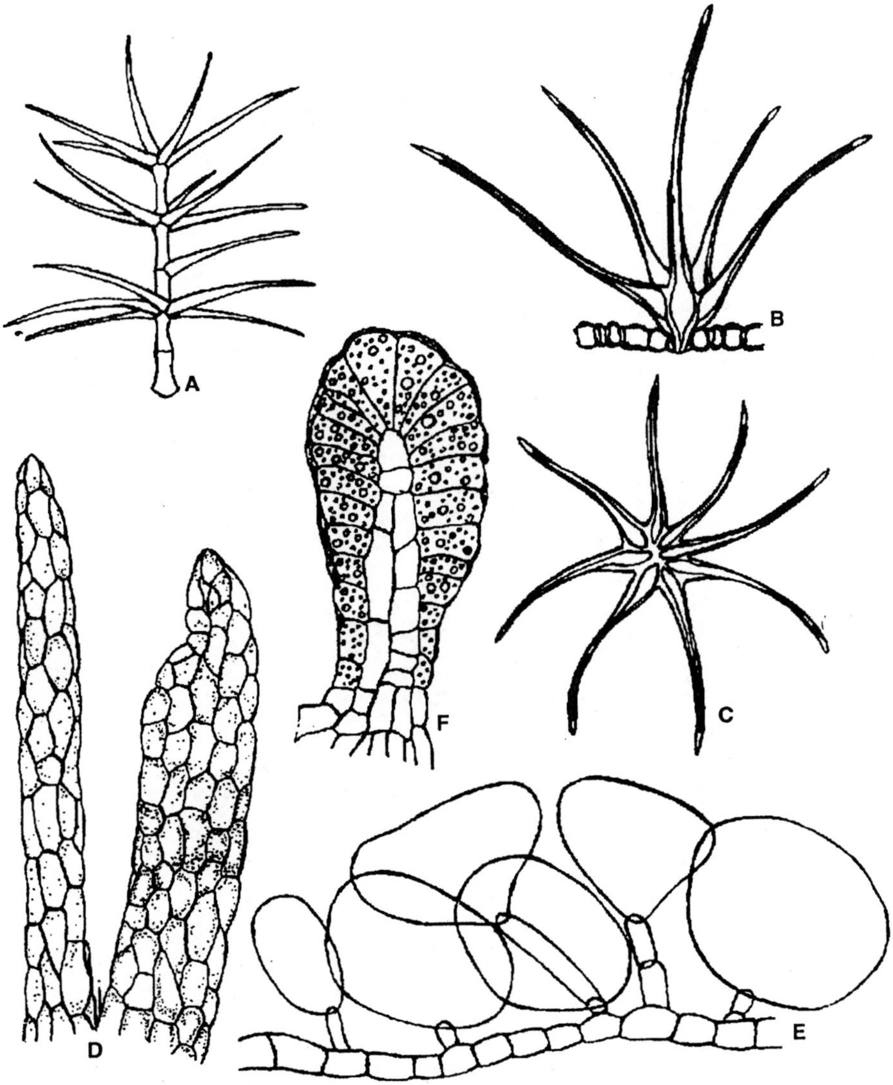

Figure 9.4: Different types of trichomes. (A) Candelabrum-like, branched, multicellular trichome from the leaf of *Verbascum*; (B) Stellate multicellular trichome from the leaf of *Styrax officinalis*, lateral view; (C) As in B, but surface view; (D) Shaggy hairs as seen at the base of the petiole of *Portulaca oleracea*; (E) Vesicular hairs of *Atriplex portulacoides*; (F) Colleter on a stipule of *Viola*.

(b) Nectar-secreting trichomes (Fig. 9.6 A): The cytoplasm at the stage of secretion is very dense and especially rich in ER. It has been suggested that vesicles mainly of ER origin are involved in the secretion of nectar,

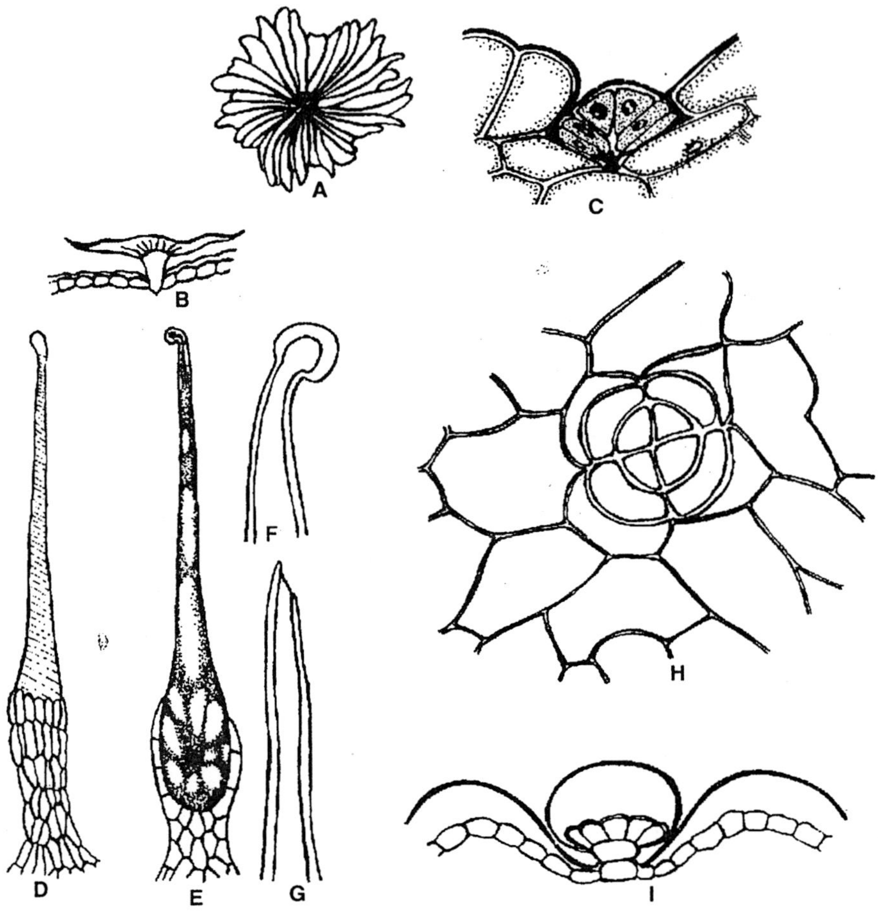

Figure 9.5: Trichomes and epidermal glands. A, B. Peltate hair of *Olea europaea*. (A) Surface view; (B) Lateral view; (C) Portion of a cross-section of the leaf of *Tamarix* showing a multicellular salt gland; D–G. Stinging hair of *Urtica dioica*; (H) Chalk gland of *Plumbago capensis* in surface view of the epidermis; (I) Portion of a cross section of the leaf of *Thymus capitatus* showing a secretory gland.

e.g., *Abutilon*. Epidermal cells not in the form of hair may also become glandular. Such secretory epidermis occurs on special outgrowths on the teeth of leaf margins, e.g., *Prunus amygdalus*.

(c) Mucilage secreting glands. These glands arise on the membranous sheath from the leaf base of *Rumex* (Fig. 9.6 C). The secreted mucilage is mainly a polysaccharide. Golgi vesicles are involved in its secretion. The extruded mucilage accumulates in spaces between the cell wall and the cuticle. The cuticle may rupture and the mucilage thus reaches the

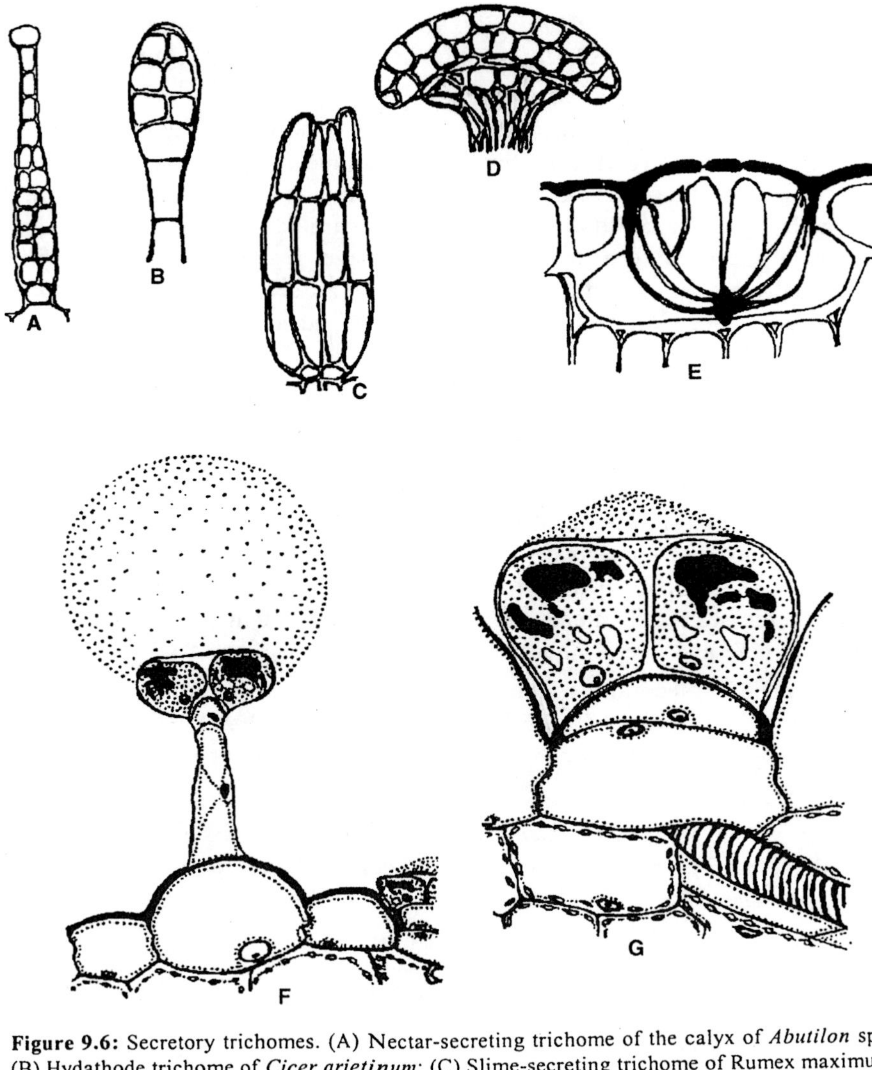

Figure 9.6: Secretory trichomes. (A) Nectar-secreting trichome of the calyx of *Abutilon* sp.; (B) Hydathode trichome of *Cicer arietinum*; (C) Slime-secreting trichome of Rumex maximus; (D) A stalked gland of *Drosophyllum*; (E) Salt gland of *Limonium latifolium*; (F, G) Stalked and sessile gland, respectively of *Pinguicula grandiflora*.

surface. Colleter is a multicellular hair having a sticky secretion (Fig. 9.4 F).

(d) Glands of carnivorous plants (Fig. 9.6 D, F, G): The capture organs of these plants are usually modified leaves. The prey — insects and often other small animals — is attracted to the capture organs by attractive colours, odour and nectar secretion. The animals are captured by the organs in various ways; by mucilages secreted by special trichomes

(*Drosophyllum, Pinguicula* (Fig. 9.6 F, G); by rapid closure of the two halves of the leaves of *Dionaea*, by attraction to pitfalls (*Nepenthes*) or by a complicated trap as in *Utricularia* ('mousetrap').

(e) Colleters, Trichomes secreting sticky substances: Glandular trichomes consisting usually of a multicellular head and a stalk which may sometimes be lacking. All external and often also the neighbouring epidermal cells have the ability to secrete. The secreted sticky substance, frequently a mixture of terpenes and mucilage, reaches the gland surface usually by rapid rupture of the cuticle. The secretion continues for a long period. Colleters appear mostly on bud scales (e.g., *Viola* (Fig. 9.4 F), *Alnus*, *Coffea* and *Rosa*).

(f) Stinging hairs: In *Urtica dioica* specialized glandular trichomes, known as, stinging hairs are found (Fig. 9.5 D–G). These possess a peculiar type of cell structure for releasing the contents of glands. These are unicellular structures with a broad calcified lower end and long, fine capillary like silicified upper end. Their base remains covered with collar-like growth of epidermal cells. When it comes in contact with the skin of an animal the tip of the hair breaks off. The sharp edge penetrates into the skin and injects the irritating cell contents, which is rich in histamine and acetyl choline.

(g) Terpene — secreting trichomes: The essential oil-secreting glands are seen in Lamiaceae (Fig. 9.5 I). These consists of a basal cell, a uniseriate stalk (one or several cells long) and a head of one or several secretory cells.

In young stems and leaves of *Mesembryanthemum crystallinum* (so called ice plant), some of the epidermal cells become greatly distend and serve as water reservoir. These are often called as ice-beads.

CHAPTER X

PERIDERM

Due to continued secondary growth, the stem gradually keeps on increasing in width and secondary tissues formed exert pressure on phloem, cortical and epidermal cells. Due to pressure, the phloem cells are pushed outwards, the cortical cells become flattened, the hypodermal cells are disturbed and become flattened tangentially and epidermis is put to maximum stretching. The pressure increases so much so that epidermis ultimately gets ruptured exposing the living cells to the external atmosphere. At this stage epidermis is replaced by a group of secondary tissues which are protective in function. The term periderm is applied to these cells.

STRUCTURE OF PERIDERM

The periderm includes the phellogen (cork cambium), the meristem that produces the periderm, phellem (commonly called cork), the protective tissue formed outward by the phellogen and phelloderm, a living parenchyma tissue, formed inward by the meristem. Death of tissues lying outside the periderm results from an insertion of nonliving cork between these tissues and the living inner tissues of the axis.

The phellogen is relatively simple in structure. In contrast to the vascular cambium it has only one form of cells. In cross section the phellogen commonly appears as a continuous tangential layer (lateral meristem) of rectangular, radially flattened cells, each with the derivatives in a radial file extending outward through the cork cells and inward through the phelloderm cells. In longitudinal section, the phellogen cells are rectangular or polygonal in outline, sometimes somewhat irregular. The protoplasts of these cells contain vacuoles of different sizes. They may also contain chloroplasts and tannins.

The cork cambium, like the vascular cambium, exhibits periods of activity and inactivity. In *Quercus* the period of activity of the cork cambium and the vascular cambium coincide. In *Robinia* two periods of phellogen activity occur during the single period of cambial activity. In *Acacia* three alternating periods of phellogen activity and inactivity occur during the year. Several factors are involved in the activity and initiation of cork cambium. In *Robinia* it is mainly active under a combination of short day and high temperature or long day and low temperature. Gibberellic acid and Napthalene acetic acid have a distinct retarding effect on initiation of cork cambium in this plant. In *Eucalyptus*, high humidity and a continuous flush of oxygen cause early initiation of cork cambium.

Phellem (cork) cells are characterized by suberization of their walls. The suberin, a fatty substance, usually occurs as a distinct lamella that covers the original primary cellulose wall.

The cork cells are suberised and thick walled, dead, compactly packed and arranged in a radial file without intercellular spaces. They lose protoplast after differentiation and become dead and remain filled with air and coloured organic matters.

Cork used commercially as bottle cork has thin walls and lumina filled with air. It is highly impervious to water and resistant to oil. It is light in weight and has thermal insulating qualities. Mature cork of this type is also a compressible, resilient tissue. The commercially valuable properties, imperviousness to water and insulating qualities, also make the cork effective as a protective layer on the plant surface.

In many species the phellem consists of cork cells and non-suberized cells called phelloid cells. Like the cork cells, the non-suberized cells may have thick or thin walls, and they may differenctiate as sclereids. In *Abies* species layers of cork cells alternate with those of cells having thick sclerified outer tangential walls. Two more common type of cork cells occur; they are: (1) Hollow, thin walled and radially widened; and (2) thick walled and radially flattened. The cells of the second type may often be filled with dark resiniferous, tanniferous substances as in *Eucalyptus*. These two types of cork cells may occur together in the same plant as in *Betula*. Here they occur in alternating layers and facilitate the cork to peel like sheets of paper.

The phelloderm cells resemble cortical parenchyma cells and may be distinguished from the latter by their position in the same radial files as the phellem cells. In certain plants the cells of phelloderm contain chloroplasts and are photosynthetic. These cells may also store starch. Sclereids and other special cells are sometimes present among the cells of phelloderm.

Development: The first periderm commonly appears during first year of growth of stem and root. The subsequent deeper periderms may be initiated later in the same year or many years later (species of *Carpinus*, *Fagus*, *Quercus*) or may never appear. The first periderm of stem originates most commonly in the subepidermal layer (*Populus, Juglans, Ulmus*) (Fig. 10.1 A).

Occasionally the first cork cambium is formed in the epidermis itself (*Solanum, Quercus, Nerium*). In stems of certain plants it has its origin in the second or third cortical layer (*Robinia, Aristolochia, Pinus*). In some species, however the first periderm arises rather deeply in the stem (*Berberis, Ribes, Vitis*) usually in the primary phloem. In most of the roots the first periderm originates in the pericycle, but it may appear near the surface as, for example, in some trees and perennial herbaceous plants in which the root cortex serves for food storage. The subsequent periderms arise successively deeper layers beneath the first and thus, eventually originate from parenchyma of secondary phloem, including ray cells.

The first phellogen is initiated either uniformly around the circumference of the axis or in localized areas and becomes continuous by a lateral spread and meristematic activity. The subsequent periderms appear most commonly as a discontinuous but overlapping layers.

The phellogen is initiated by periclinal divisions and produces the phellem and the phelloderm by periclinal divisions (Fig. 10.1 A, B). The phellogen keeps pace with the increase in circumference of the axis by periodic divisions of its cells in the radial anticlinal plane.

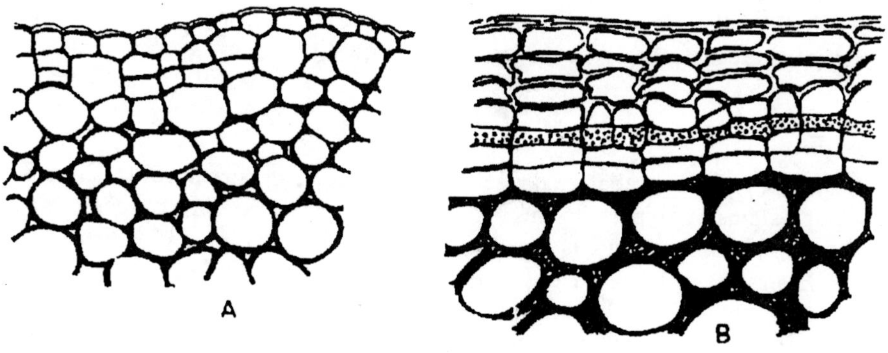

Figure 10.1: Origin of periderm in *Pelargonium* stem as seen in cross sections. (A) Periclinal divisions in subepidermal layer; (B) Periderm is established.

WOUND PERIDERM

Wounding induces a series of metabolic events and related cytological responses that lead, under favourable conditions, to a complete closure of the wound. Natural wound periderms are basically alike in method of origin and growth and may have the same cellular components. The natural periderm develops beneath a surface which is effectively sealed by the cutinized epidermis. Correspondingly, the formation of wound periderm is preceded by a sealing of the newly exposed surface by scar (cicatrice) tissue. This tissue includes dead (necrosed) cells on the surface and living cells beneath, which become suberized and lignified and form the so-called closing layer. The wound phellogen arises beneath the closing layer and, when cork develops, the dead scar tissue is carried outward. Successful development of wound periderm is important in horticultural practice when plant parts used for propagation must be cut (e.g., potato tubers, sweet potato roots). Environmental conditions markedly influence the development of wound periderm.

Plant taxa vary with regard to the histological aspects of wound healing, as they do in details of natural development of the protective tissue. In general Monocotyledons are less responsive to wounding than the Dicotyledons. In Dicotyledons and certain Monocotyledons (Liliales, Araceae, Pandanaceae) healing includes formation of both closing layer and wound periderm. In other Monocotyledons no wound periderm is detectable. Among these, the Zingiberales produce a slightly suberized closing layer whereas the Arecaceae and the Poaceae form a lignified closing layer.

A wound reaction occurs when the periderm is peeled off to the living cells underneath. The newly exposed cells die, and a new periderm arises below them. This reaction is utilized in the production of commercial cork from the cork of oak. The first cork, which is of inferior quality, is removed up to the phellogen and the new phellogen developing beneath the scar tissue produces massive cork of superior quality.

POLYDERM

A special type of protective tissue called polyderm occurs in roots and underground stems of Hypericaceae, Myrtaceae, Onagraceae and Rosaceae. It consists of alternating tissue layers, layers, one cell deep of partly suberized cells and layers several cells deep of non-suberized cells. The polyderm may become twenty or more layers in total thickness, but only outermost layers are dead. In the living part, the non-suberized cells function as storage cells.

COMMERCIAL CORK

Commercial cork is obtained from the phellem of oak — *Quercus suber* — of family Fagaceae. It grows in Mediterranean region. In the stem of this plant the first cork cambium forms in epidermis. It may remain on the plant indefinitely but in order to obtain commercial cork, this first formed periderm is removed when the tree is about 20 years old and about 40 cm in diameter. After the first bark/periderm is removed, the exposed cells of the primary cortex and secondary cortex dry out and die. A new cork cambium is formed a few millimeters within the cortex, below the first formed cambium. This subsequent cambium produces cork more rapidly and in about 10 years a cork layer, thick enough to be of commercial value, is produced. This cork is of better value than the first formed one, which has almost no commercial value. The features that give commercial value to cork are its imperviousness to gases and liquids and its strength, elasticity and lightness.

RHYTIDOME

As a tree grows older, periderms often arise at successively greater depths and thus cause an accumulation of dead tissues on the surface of stem and root. This dead part of the bark composed of layers of tissues isolated by the periderms and layers of no longer growing periderms is called rhytidome. Rhytidome thus constitutes the outer bark and is especially well developed in older stems and roots of trees. In shrubs, early exfoliation of older bark in common and precludes an accumulation of a thick rhytidome.

BARK

Bark is a non-technical term applied to all tissues outside the vascular cambium or the xylem. In older trees bark may be divided into dead outer bark and living inner bark, which consists of secondary phloem.

Two types of barks are found-the ring bark and the scale bark. If the bark is found in the form of concentric rings surrounding the entire stem, it is called ring bark, e.g., *Vitis, Lonicera, Betula, Clematis, Mangifera*. If the bark is found in the form of separate strips or patches it is called scale bark, e.g., *Ficus, Pinus, Pyrus, Psidium guajava*.

LENTICELS

A lenticel may be defined as a limited part of the periderm in which the pellogen is more active than elsewhere and produces a tissue that, in contrast to the phellem, has numerous intercellular spaces. The lenticel phellogen itself also has intercellular spaces. Because of this relatively open arrangement of cells, the lenticels are regarded as structures permitting the entry of air through the periderm.

Lenticels are usual components of periderm of stems and roots. Outwardly, a lenticel often appears as a vertically or horzontally elongated mass of loose cells that protrudes above the surface through a fissure in the periderm. Lenticels vary in size from structures barely visible without magnification to those 1 cm and more in length. They occur singly or in rows. Vertical rows of lenticels frequently occur opposite the wide vascular rays.

The phellogen of a lenticel is continuous with that of the corky periderm but usually bends inwards so that it appears more deeply situated. The loose tissue formed by the lenticel phellogen toward the outside is complementary or filling tissue, the tissue formed toward the inside is the phelloderm.

Two types of complementary tissue are distinguished: that in which the connection between the cells is relatively strong, as for example in *Sambucus nigra, Salix* and *Ginkgo* and that in which the cells have almost no attachment between them, giving the tissue a powder-like appearance as for example, in the stems of *Pyrus, Prunus* and *Robinia* and in the roots of *Morus*. In the latter type the cells of the complementary tissue are held in place by closing layers (Fig. 10.2 B).

The first lenticels frequently appear under stomata. The parenchyma cells beneath a stoma undergo some preparatory divisions; then a phellogen pushes the overlying cells outward and rupture the epidermis (Fig. 10.2 A).

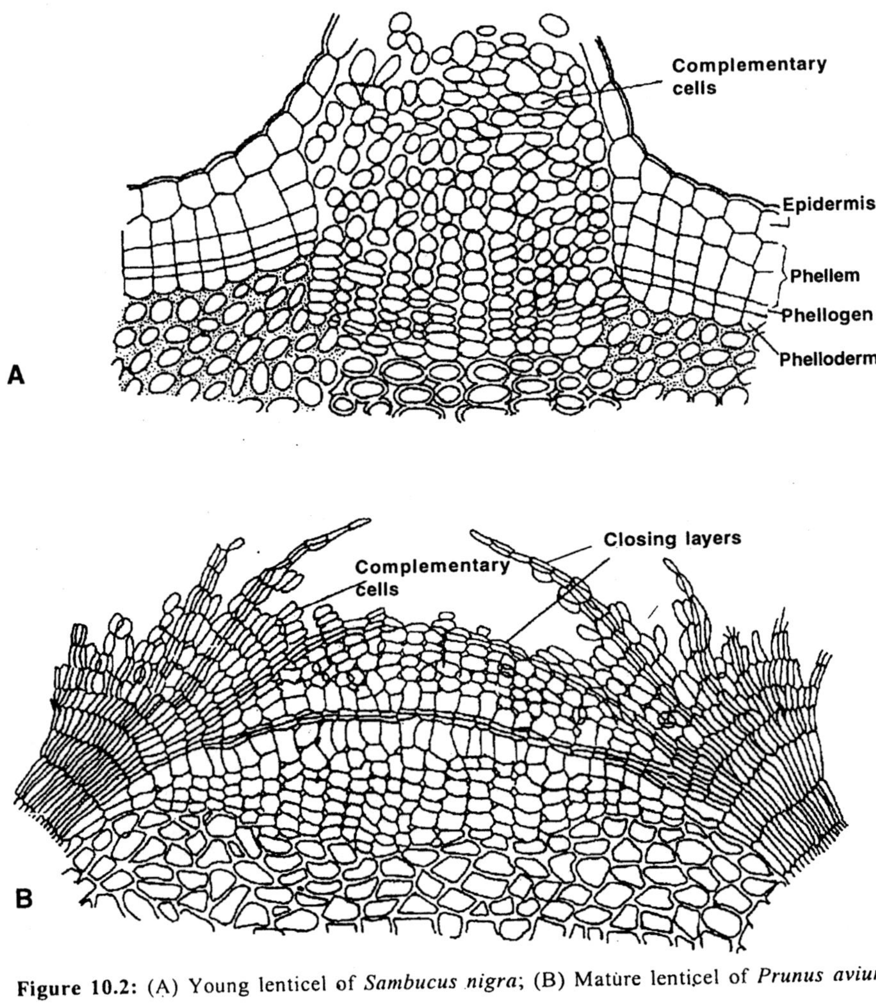

Figure 10.2: (A) Young lenticel of *Sambucus nigra*; (B) Mature lenticel of *Prunus avium*.

Lenticels are maintained as long as the periderm continue to grow. Deeper periderms also have lenticels.

In Dicotyledons, three structural types of lenticels are recognized. In the first type, the lenticel has a filling tissue composed of suberized cells. e.g., *Magnolia, Salix, Malus* etc. In the second type lenticel consists of a mass of more or less loosely structured non-suberized filling tissue, succeeded at the end of the season by a more compact layer of suberized cells. e.g., *Quercus, Sambucus, Tilia* etc. In the third type the filling tissue is layered because loose non-suberized tissue regularly alternates with compact suberized tissue, e.g., *Betula, Fagus, Prunus* etc.

PROTECTIVE TISSUES OF MONOCOTYLEDONS

In herbaceous monocots the epidermis, which has a thick cuticle, is the only external protection on the plant axis. When the epidermis is ruptured the cortical cells below become suberized as in typical cork cells (Poaceae, Juncaceae, Typhaceae).

In perennial monocotyledons different types of protective tissues are formed on the stems. In palm, *Roystonia*, the bark is white and smooth and possesses a hard periderm, which is similar to the Dicotyledons. It remains on the stem through out the life of the plant. In *Curcuma, Cordyline* and many palms, a special kind of protective tissue is formed. This tissue is formed by the secondary activity of a storied meristem which develops in the outer cortex. The initials of this meristem undergoes three to eight periclinal divisions and gives rise to radially arranged layers of cells, which become suberized. Cork of this type is called storied cork. The initials here do not form a regular uninterrupted cylinder.

In species of *Dracaena, Aloe arborescens* and *Yucca*, the secondary protective tissue developed in them are almost similar to those of Dicotyledons, which are produced by a typical cork cambium.

CHAPTER XI

ROOT

The root constitutes the lower portion of the plant axis and it usually develops below the soil surface, although there are roots that grow in air. The differences in the development and arrangement of the primary tissues in these two organs are many. The histogenesis of the epidermis of the root differs from that of stem. In seed plants the primary xylem of the root is exarch and that in the stem is endarch. The xylem and phloem strands in the root do not form common bundles but are arranged alternately, while in stem vascular bundles are collateral, bicollateral or amphivasal. Roots bear no appendages that are comparable to the leaves of the stem; roots are devoid of stomata and their branches originate in the relatively mature tissue of the pericycle in contrast to the stem where the branches originate from the apical meristem. Roots also possess a root cap which has no parallel in stems.

The radicle present in the seed consists of root meristem and it gives rise to the first root on germination of the seed. In Gymnosperms and Dicotyledons this root develops into the tap root, known as seminal root, with its branches. In Monocotyledons first root usually dies out early in the growth of the plant and the root system of the mature plant consists of adventitious roots. The roots of mature Monocotyledons are usually adventitious and develop from the stem. They may branch several times or they may be unbranched. Generally such roots do not develop secondary thickening.

A tap root system generally penetrates the soil more deeply that the fibrous roots do, but the later bind more firmly the superficial layers of the soil. The tap root and its larger branches undergo secondary growth but the small absorbing roots remain in the primary state and are often ephemeral.

The root system is concerned with anchorage, absorption, storage and conduction.

PRIMARY STRUCTURE

Root Cap

The root cap is situated at the tip of roots, it protects the root promeristem and aids the penetration of the growing root into the soil. The root cap consists of living parenchyma cells derived from the apcial meristem by divisions contributing cells away from the apex. This part of the apical meristem is called calyptrogen. As new cells are produced cells on the periphery of the root cap are sloughed off. In *Avena* the time interval between the origin of root cap cell and its shedding from the cap was calculated to be 5 to 6 days. Environmental conditions affect the development of the root cap. When plants normally growing in soil are raised in a water culture their roots may fail to develop root caps. Aquatic plants, however, often have particulary large root caps.

The root cap has long been regarded as an organ controlling georeaction of the root. When roots normally growing downward are placed horizontally, the subsequent growth in the elongation zone occurs along a curve until the normal downward orientation is attained (Fig. 11.1). When the tip is excised or the root cap removed, growth of the root is not stopped but the geotropic reaction is lost until the root cap is regenerated.

The site of perception of gravity in the root cap is the central column of cells in which starch-containing amyloplasts act as statoliths, that is gravity sensors. In vertically growing roots the amyloplasts are sedimented on the distal horizontal walls of the cells in the central column (Fig. 11.2 A). A few minutes after the root is placed horizontally the amyloplasts slide towards the longitudinal walls directed down ward (Fig. 11.2 B). With in some 24 hours the root assumes its normal vertical position by growing along a curve and the amyloplasts return to their previous location next to horizontal walls.

Epidermis: The epidermal cells of roots are thin-walled and are usually devoid of cuticle. The epidermis of roots is usually uniseriate. However in the aerial roots of tropical Orchidaceae and epiphytic Araceae the epidermis is multiseriate and it is specialized to form a velamen. The velamen consists of nonliving cells, compactly arranged and often bearing secondary wall thickenings in the form of numerous narrow strips. The velamen is commonly interpreted as an absorptive tissue, but some physiological studies on orchid velamen indicate that the principal roles of this tissue are mechanical protection and reduction in the loss of water from the cortex.

In young roots the epidermis is specialized as an absorbing tissue and usually bears root hairs, which are tubular extensions of epidermal cells.

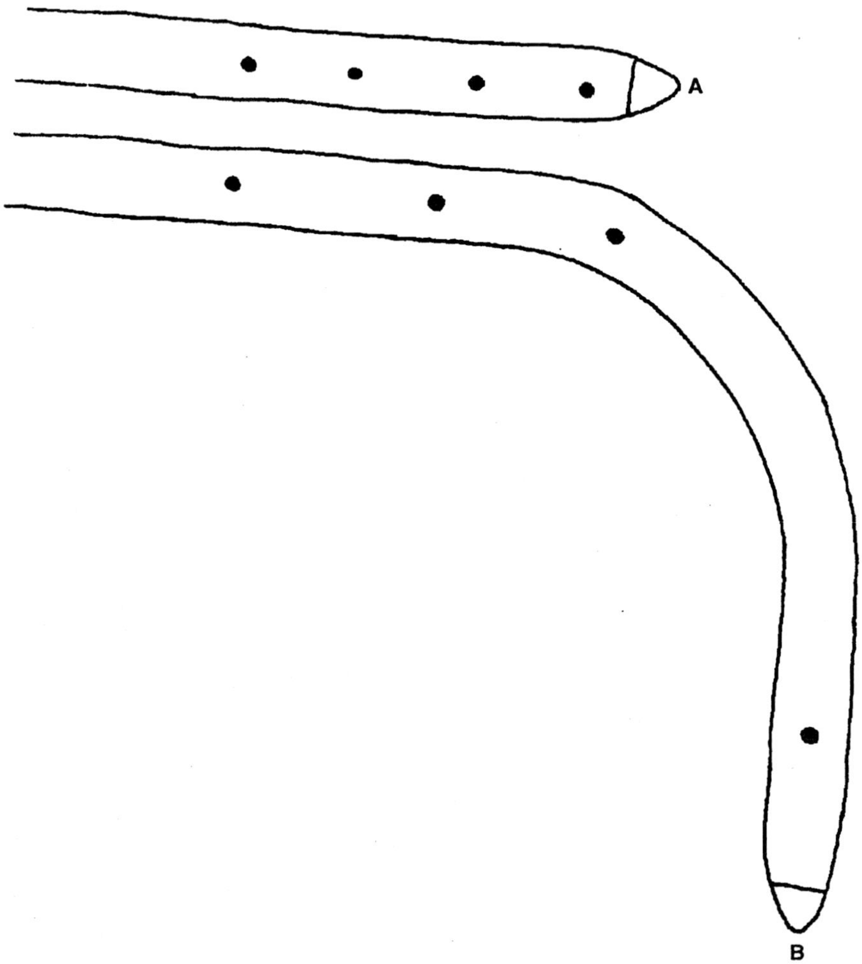

Figure 11.1: Diagram illustrating response of root tip to gravity. (A) Root tip of root normally growing down-ward is placed horizontally; (B) 23 hours later the root tip grows along a curve and returns to a vertical orientation with regard to the soil.

The region of root hairs is usually restricted to one or a few centimeters from the root apex, where elongation has ceased. Root hairs are absent close to the apical meristem and they usually die and dry out on the more mature portions of the root. Certain herbaceous plants, especially water plants lack root hairs. Plants that usually grow in soil and which produce root hairs fail to do so when they are grown in water.

Epidermal cells of young roots and root hairs of some genera secrete fibrillar mucilage forming a mucilaginous layer. In plants grown on soil this layer was found to be colonized by bacteria.

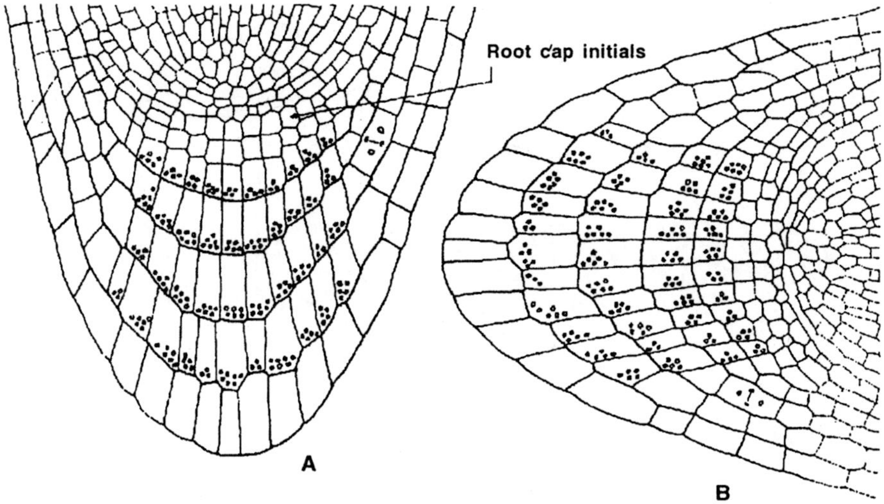

Figure 11.2: Diagram illustrating response of amyloplasts with starch (small circles), the statoliths, to gravity. (A) Amyloplasts are sedimented on distal horizontal walls in root cap of root growing vertically downward; (B) In the same root placed horizontally, the amyloplasts slide toward normally vertical walls that are now parallel with the soil. This change in position is involed in the georeaction shown in Figure 11.1.

Cortex: Cortex of the root consists of mainly parenchyma cells. In many monocotyledons cortex consists of much sclerenchyma in addition to parenchyma. In aquatic plants the cortex is largely aerenchymatous enclosing air cavities. The air spaces may have either a schizogenous or a lysigenous origin or both.

Cortical cells are usually devoid of plastids but are accumulating various amounts of starch. Some aerial climbers such as *Tinospora* have got photosynthetic or assimilatory roots, and the chloroplasts are found in their cortical cells. The root cortex is usually wider than the stem cortex and therefore it plays a larger role in storage. The arrangement of cells may also vary, for example, in *Avena sativa,* cortical cells are arranged in distinct radial rows. In *Commelina* the cells of inner cortex are arranged in the form of concentric rings. In most Monocotyledons, some Dicotyledons and Gymnosperms, the sub-epidermal or outer few cortical layers differentiate into protective tissue called exodermis. The walls of these cells become suberized. Also lignin is deposited in the walls of these cells and, in certain cases, casparian strips have also been distinguished.

The innermost layer of the cortex is differentiated as an endodermis. In the absorbing region of the root the endodermal cell wall contains suberin in a band like region extending completely around the cell within the radial and transverse walls (Fig. 11.3 A, B). This band, called casparian strip, is not merely a wall thickening but an integral part of the primary wall, for the

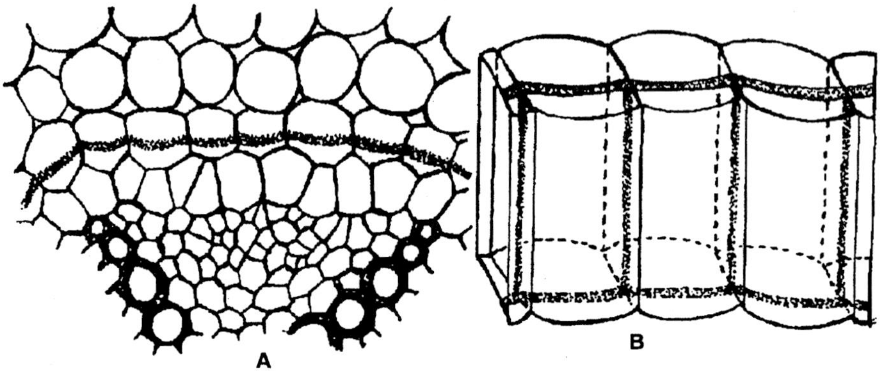

Figure 11.3: Structure of endodermis. (A) Cross section of part of *Convolvulus arvensis* showing position of endodermis with regard to xylem and phloem; (B) Diagram of three connected endodermal cells oriented as they are in A; Casparian strip occurs in transverse and radial walls (that is, in all anticlinal walls) but is absent in tangential walls.

deposition of suberin is continuous across the middle lamella. In the Electron microscope the casparian strip appears as a slightly thickened, homogenous electron opaque wall region. If plasmolysis is brought about in endodermal cells the protoplast withdraws from the tangential walls but remains attached to the casparian strips. This type of plasmolysis is called band plasmolysis. Plasmodesmata occur in all wall portions of the endodermal cells except for the casparian strips in which the plasmalemma adheres. In transpiring plants the solute moves through the cortex in the cell walls (free space of apoplastic pathway). It is blocked in its apoplastic progress by casparian strip and is forced to pass through the protoplast of the endodermal cells (symplastic pathway), and resume the apoplastic course in the vascular cylinder until the solute enters tracheary cells. Thus, the movement of solutes is subjected to the control of the differentially permeable protoplast membranes within the endodermis.

In roots with considerable amount of secondary growth the endodermis is usually cast off together with the cortex, but in those remaining in primary state (commonly root of monocotyledons) it often develops thick secondary walls. These walls typically consist of a suberin lamella covered by layers of lignified cellulose. Formation of the secondary wall may be delayed in endodermal cells opposite the xylem. Such thin walled cells (without casparian strips) in an otherwise thick-walled endodermis are called passage cells.

The endodermis is usually uniseriate, but in many Asteraceae certain young endormal cells divide tangentially, and schizogenous secretory canals develop in the two-layered zones.

The exact and basic function of endodermis is still not clear. Some functions, on the basis of its relation to water and vascular tissues have been attributed to it. (i) It is considered as a water-tight jacket between xylem and cortex, (ii) It acts like an air dam that prevents diffusion of air into xylem which would otherwise get clogged, (iii) It is a protective layer, sometimes referred to as accessory epidermis, (iv) It regulates and maintains root pressure, (v) It acts as a water dam due to presence of compactly set thickened cells, (vi) It provides regulated and systematised flow of water through its passage cells, (vii) The thin walled cells retain power of division, as such, the lateral branch and adventitious buds often originate through this layer.

In many plants the walls of the cells of the outer subepidermal layers of the cortex become suberized. In this way the exodermis, a protective tissue, is formed. This exodermis is similar in structure and cytochemical characteristics to the endodermis. Its cells may have casparian strips, but more commonly they appear to have a suberin lamella covered by a cellulose wall. The exodermis contains either one kind of cell or short and long cells. The exodermis may be several layers in thickness. The cortex is sclerified beneath the exodermis in Poaceae, Arecaceae and Cyperaceae. In *Phoenix* the exodermis is fibrous.

VASCULAR CYLINDER

The primary vascular tissue is surrounded by a region of cells which is termed the pericycle. The pericycle is commonly one layer of cells in thickness but may be multiseriate. The pericycle may be composed entirely of parenchyma or it may contain sclerenchyma. It may be interrupted by elements of protoxylem. The primordia of the lateral roots in all spermatophytes and the phellogen and portions of the vascular cambium in the Dicotyledons develop form the pericycle. Schizogenous secretary canals occur in the pericycle of the Apiaceae. In roots of certain water plants and parasites the pericycle is absent.

The xylem usually forms a solid core with ridge like projections, seen as radial files of cells in transections, extending towards the pericycle. Strands of phloem alternate with the xylem ridges. The number of xylem groups varies in different species and in relation to this variation the roots are called monarch, diarch, triarch, tetrarch etc. or polyarch. *Trapa natans* has a thin root which is monarch. Diarch roots (Fig. 11.4 A) are found in *Lycopersicon, Nicotiana, Beta, Raphanus, Daucus* and *Linum*. In *Pisum* the root is triarch (Fig. 11.4 B), while in *Cicer, Vicia, Ranunculus* and *Gossypium*, it is tetrarch (Fig. 11.4 C). Polyarch arrangement (Fig. 11.4 D) is characteristic of the adventitious roots of Monocotyledons. In Monocotyledones the number of xylem strands in the seminal root is small, as in dicotyledonous roots, but the adventitious roots are polyarch and the number of strands in the Arecaceae and the Pandanaceae may be 100 or more.

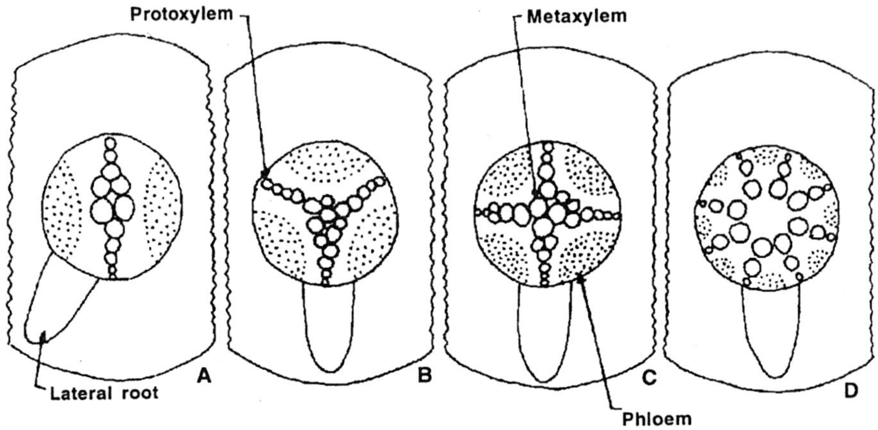

Figure 11.4: Different patterns formed by primary xylem in cross sections of roots and position of lateral root with regard to xylem and phloem of main root. (A) Diarch; (B) triarch; (C) Tetrarch; (D) Polyarch.

The primary vascular structure of monocotyledon roots is highly variable and often complex. In some, the center is occupied by a single metaxylem vessel, (such as *Triticum*), in others a circle of such vessels surrounds a pith (*Zea* and *Iris*). The number of protoxylem points may exceed a hundred in large palm roots and in some monocotyledons, phloem strands are scattered through out the xylem core.

The tracheary elements in the root mature centripetally and therefore the xylem is exarch, i.e., the protoxylem is situated on the outer side of the metaxylem. The differentiation of the phloem is also cetripetal so that the protophloem is closest to the pericycle, while the metaphloem is closest to the axis of the root.

Companion cells are characteristic of the metaphloem but may be absent in protophloem. In roots having secondary growth the cells located between the xylem and the phloem eventually function as a vascular cambium. In roots without secondary growth the cells in the same location mature as parenchyma or sclerenchyma cells.

The primary phloem of the roots of most plants does not contain fibres but in certain plants such as those of the Papilionaceae, Malvaceae and Annonaceae fibres are found in the primary phloem.

MYCORRHIZAE

The epidermis and cortex of roots of many plants are often associated with soil fungi. This association between fungal hyphae and the young roots of higher plants is known as mycorrhiza (Greek: *Mykes* = mushroom, *rhiza* =

root). Usually this is symbiotic, both the higher plant and the fungus derive benefit from the association.

Mycorrhizae are classified according to the relation of the fungus to the cortical cells of the host. The two principal forms of mycorrhizae are (i) Ectomycorrhizae, in which the fungus envelops the entire root tip with a dense sheath called hyphal mantle and enters intercellular spaces, (ii) Endomycorrhizae, in which the fungus forms an inconspicuous mantle but invades the interior of the cells.

In ectomycorrhizae fungus remains predominantly on the outside of the root, with a small amount of penetration between the outer cortical cells. Ectomycorrhizal roots are short, branched and appear swollen. On the surface of the root, the fungal mycelium forms a thick, dense sheath or mantle. From the sheath, hyphae penetrate into the intercellular spaces of the epidermis and cortex. No intracellular penetration into epidermal or cortical cells occurs, but an extensive network called the "harting net" is formed between these cells. In ectomycorrhizal roots the development of root hairs is depressed and the volumes of apical meristem and root cap may be reduced. Ectomycorrhizae occur in many plants, e.g., *Pinus, Picea, Abies, Cedrus, Larix, Quercus, Castanea, Fagus, Betula, Populus, Carya, Salix* and *Eucalyptus*.

In the endomycorrhizae, there is little surface mycelium but an extensive network within the root. The Endomycorrhizal roots are similar to uninfected roots in form but are darker in colour. The endomycorrhizae are divided into two groups — those involving septate fungi and those involving non-septate fungi. (i) Septate fungi are rare and forming associations only with orchids and some members of the order Ericales. The orchids are extremely dependent on their mycorrhizal associations. If proper fungi were not invaded, the seedlings would die. So they are essential for successful plant growth. Endomycorrhizae occur in the Orchidaceae, Ericaceae, *Acer, Liriodendron*, *Ornithogalum umbellatum* etc.

The endomycorrhizae involving nonseptate fungi are the common type present in most of the terrestrial plants. After penetrating the host produces characteristic vesicles and arbuscules. So these are called VAM (Vesicular-Arbuscular Mycorrhizae). In the soil the fungus extends as a thin mycelium and penetrates either into the cell or between the cells, when a hypha contacts a suitable root. They do not break either the plasmalemma or the tonoplast of the host while they extend into the cell. The hypha usually coils extensively, creating large surface area that produces contact with the cell. Other hyphae ramify throughout the cortex, pushing the cells slightly apart. Some of these hyphae swell, forming the vesicles for which they are named as VA mycorrhizae. In cortex, one of the hyphae invade the cells, they ramify and dichotomize, forming an arbuscule that looks as a miniature tree. The hyphae of the arbuscules fill with phosphate granules which are absorbed by the plant. Once the arbuscule has completed its transfer of phosphate to the plant, it collapses and the plant cell returns to a state similar to that of an uninvaded cell. While this happens, the plant cell responds by increasing the amount of cytoplasm it has. The protoplastids become chromoplasts. They do not accumulate starch, as the sugars are being converted to the fungi. Plant also responds by increasing the amount of vascular tissue in the area and increasing the lignin content. The fungus never penetrates the endodermis and stele.

The chief role of mycorrhizal fungi appears to be conversion of minerals of the soil and of the decaying organic material into forms accessible to the host. The host is presumably secreting sugars, amino acids and other organic materials, making them available to the fungus. Mycorrhizae also make the host plant less susceptible to drought. Some mycorrhizae increase the resistance of the host to pathogenic infections.

Root Nodules: Roots may have an association with bacteria that is beneficial to the plant. The association leads to a development of root nodules, a phenomenon particularly characteristic of Fabaceae. The bacteria capable of infecting the roots of legumes and stimulating nodule development belong to several strains (sometimes given the status of species) of the genus *Rhizobium*. Each strain of the bacterium is able to infect the roots of only a particular group of legumes, or in the case of the root nodule forming bacteria of soyabean, only one species. The bacteria are free living in the soil, but fix nitrogen only in symbiotic association with the host legume. The seedling roots may be infected by *Rhizobium* at any time after root hairs are present on them. The bacteria in the soil are attracted by secretions from the root, especially of tryptophan, which they convert to Indole Acetic Acid (IAA). The presence of this auxin may cause a root hair to grow around the bacteria nearby. The bacteria invade the root, chiefly through the root hairs, and in multiplying, form an infection thread by becoming enclosed in a sheath of gum like material. The thread penetrates deeply into the root and induces a proliferation of the inner cortical cells. This proliferation, superificially resembling a branch root primordium becomes root nodule. The mature nodule has a characteristic shape, depending upon the species of legume; it consists of an outer layer of small cells containing a vascular trace, and an inner region of much larger cells which become vacuolate and are filled with bacteria. The bacteria fix atmospheric nitrogen, using carbohydrates from the host legume as an energy source for the reduction of N_2 to NH_3 by the enzyme Nitrogenase. After 8–10 weeks the nodule begins to break down, and eventually falls from the root and disintegrates. The bacteria are then released into the soils as free-living coccoid cells.

Non-leguminous plants which show nodulation are *Alnus, Casuarina, Myrica* and *Cercopus*.

ROOT DEVELOPMENT

Root Apical Meristem (RAM) is a complex structure. To understand its development it is better to analyze simple systems. In this view point, *Arabidopsis* root is one of the ideal systems for the study of plant organogenesis. Using molecular genetic approaches, rapid progress has been made in identifying some of the pathways regulating root development. The primary root grows mainly by cell division within its meristematic zone. Newly produced cells elongate the root that was built during embryogenesis. When cells leave the meristem they are incorporated into the zone of elongation. Here they elongate along the longitudinal axis of the root. Cell elongation also contributes to the root growth. Finally, the cells enter the zone of differentiation, where they begin to produce the features that will later characterize them.

THEORIES RELATING TO ROOT APEX ORGANISATION

Several theories have been proposed to describe the root apex organization.

(a) Apical cell theory: Nageli (1878) proposed that a single tetrahedral apical cell in the root apices brings about growth. It has three cutting faces for giving rise to the tissues of the body of the root. The root cap is derived from its base. But such a type of development is not found in Angiosperms.

(b) Histogen theory: Hanstein (1968) put forth Histogen theory proposing that in root apex regions, three distinct histogens are found. The dermatogen that gives rise to epidermis, the periblem to the root cortex and the plerome to the vascular cylinder. Haberlandt (1914) proposed the terms protoderm, ground meristem and procambium, respectively for these histogens.

(c) Korper-Kappe theory: Schuepp (1917) proposed that the cells at the root apex divide in two planes. The zone with inverted 'T' type of division was referred as Korper (cap) whereas other with straight 'T' as kappe (body). The theory basically recalls the Tunica corpus theory concept of shoot apex and similarly, it also fails to explain the differential behaviour in different species.

Organization of the RAM

Developmental zones: Primary root growth occurs at the root apical meristem (RAM). The growing root tip of *Arabidopsis* is organized into a series of concentric cell layers, comprising an outer epidermis, and inner cortical, endodermal and pericycle layers surrounding the vascular tissue in the middle of the root. In these layers the number and arrangement of cells is invariant.

The root tip is also divided into several morphological zones along the longitudinal axis.

(a) The most distal part of the root tip is the root cap. This root cap protects the root as it pushes through the soil.
(b) Next to it is the zone of root apical meristem (RAM). This is also known as zone of cell division.
(c) Behind the RAM is the zone of elongation (where cell division caeses and there is rapid cell growth by elongation).
(d) Lastly the zone of differentiation, where cells assume their final fate. Cells in this zone are provided by the root hairs.

Structure of RAM

Root apical meristems (RAM) in dicotyledonous plants have two organizational schemes; closed (with highly organized) and open (tiers lacking or disorganized). These schemes are commonly believed to remain unchanged during the growth of the root. In many root meristems, cell files can be traced as a small group of initials. Often these initials are organized in a tiered arrangement, with each tier-giving rise to particular tissues. In radish, tier 1 produces root cap and epidermis, tier 2 cortex and tier 3 vasculature. In maize, tier 1 produces only root cap, tier 2 epidermis and cortex, tier 3 vasculature. Other plants show no tiered arrangement or only a single initial layer. *Arabidopsis* roots are extremely small and simple. The radial pattern consists of concentric arrangements of 6 different tissues, each a single cell layer thick.

In *Arabidopsis*, the majority of the RAM is made up of proliferating cells and it is divided into two sub-functional zones. The base of the RAM is called promeristem. This promeristem comprises three tiers of initial cells that give rise to root. The lower tier comprises a central group of 12 cells that give rise to the tip of the root cap. This tip of the root cap is called columella. The columella initials are surrounded by a ring of 16 cells, which give rise to the lateral root cap and epidermis. The middle tier of cells consists of central group of four cells known as the quiscent centre (QC). QC cells have a characteristic structure. QC acts to maintain the initials in an undifferentiated state. This strategy ensures that after each initial cell division, one daughter cell is disconnected from the QC and allowed to differentiate. The quiscent centre contains RAM's stem cells and hence it is equivalent to the central zone of SAM. This is surrounded by a ring of 8 endodermal and cortical initials, which give rise to cortex and endodermis. The cells in the upper tier give rise to pericycle and vascular bundles.

The pericycle is made up of cells that can initiate the formation of lateral roots. All the cells in the pericycle appear to be capable of lateral root initiation, but under normal circumstances only the pericycle cells nearest to the internal xylem poles perform this function. In the center of the root, the vascular tissue has bilateral symmetry with xylem on the axis of symmetry and phloem on both sides of it.

Specification of Cell Fates and Tissue Differentiation in RAM

The cell fate and tissue differentiation in RAM of *Arabidopsis* depends upon the cell lineage, cell-to-cell interaction, position of the initial cell and regulation of the quiescent centre.

Cell lineage analysis has shown that the organization of the root is generated by a stereotypical series of cell divisions in the promeristem. Here, all these patterns of cell division are invariant.

Although pattern of cell division in the RAM of *Arabidopsis* may be invariant cell lineages, this does not mean that the fate of each cell is determined from the outset of the development. In *Arabidopsis* roots cell-to-cell interaction is also important in determining cell fate.

Laser ablation studies in the RAM of *Arabidopsis* indicate that position rather than lineage is the important factor for determining cell fate. For example, (1) If a cortical/endodermal initial cell in the second tier of the promeristem normally give rise to daughter cells that divide to produce cells of the cortex and endodermis. (2) If the cortical/endodermal initial is ablated with a laser beam, the adjacent pericycle initial undergoes an unexpected division and replaces it, which assume the pattern of cell division characteristic of cortical/endoderm initial and give rise cortical and endodermal derivatives. Thus the cell changes fate from pericycle to cortical initial when it's position is altered. (3) If one daughter cell is ablated, the underlying cortical/endodermal initial undergoes an additional division to generate a new one. (4) If three cortical cells are ablated, the cortical endodermal initial cell underlying the middle daughter cell is no longer able to produce any more daughter cells that divide. Therefore this experiment reveals that signals from the cortical daughter cell controls the developmental potency of the underlying initial cells.

The quiescent centre also regulates the differentiation of neighbouring cells. If one or two QC cells were ablated, the columella initials that were no longer in direct contact with a QC cell ceased dividing and underwent differentiation as a columella cell. Similarly cortical initial cells that were no longer in direct contact with the QC cell behaved as cortical daughter cells and underwent an asymmetric division to generate cortical and endodermal cell files. Therefore, direct contact with the QC inhibits the differentiation of initial cells in the RAM (van den Berg, 1997).

Molecular basis of RAM development

The developmental potency of endodermal/cortical initial cells is found to be interfered by several *Arabidopsis* mutants like *short root mutant*, *scarecrow mutant* and *pinocchio mutant.*

1. In short root mutant, the endodermal and cortical initial cells fail to divide and produce a single file of cells that differentiates as cortex. Here endodermis is absent.
2. In scarecrow mutant, the endodermal and cortical initial cells fail to divide and produce a single tier of cells that is an intermediate in character between the cortex and endodermis.
3. In pinocchio mutant, the initial cells fail to divide but here endodermis is present while the cortex is missing.

Most laser ablation experiments result in normal development of the root. Because, when the ablated cells are crushed, which are replaced by the divisions of adjacent cell. This shows that cell-cell interactions (or) cell-cell signaling plays a very important role in the normal development of root. The *fass mutant* confirms this.

1. The *fass mutant* consists of fat roots with more number of cells than extra cell layers. Therefore this *fass mutant* can partially overcome the deficiency of *scarecrow mutant* phenotype, as the *scarecrow mutant* is deficient in the generation of extra cell layers. When these two mutants are combined, the double *fass* + *scarecrow mutant* has both cortex and endodermis.
2. When the double fass mutant and short root mutant was combined, even though extra layers are formed, they still lack endodermis. By this it is clear that in short root mutant, the defect must be in molecular pathway that determines the identity of the endodermal cells but not the lack of extra cell layers.
3. Another root mutant, where their phenotypes are analogous to the *shoot meristemless* and effects the proliferation in RAM. The corresponding genes are named as:

   Root meristemless 1
   Root meristemless 2.

These genes do not show any defect in SAM proliferation, where as the cell proliferation is abolished in short root mutant where endodermis is absent. This indicates that the cell proliferation of RAM is linked with the ability of the endodermal/cortical initials, which divide, and produce endodermis (or) in short root mutant the endodermal initials fail to divide and do not give rise to endodermis. As the cell proliferation of RAM is linked with the ability of the endodermal initials, no cell proliferation of RAM could be seen in short root mutant.

ROOT HAIR — BIOLOGY AND DEVELOPMENT

Root hairs are tubular-shaped, tip-growing cells that arise from root epidermal cells known as trichoblasts. Root hairs are thought to increase the absorption capacity of the root. Root hairs are normally 0.1–1.0 cm long. They can be extremely numerous with more than 20,000 hairs/cm^2 of root and billions per plant increasing a vast surface area for absorption. Root-hairs are short lived and no hairs are seen on older part of the root. The root hair zone is therefore extremely important for water and mineral nutrient uptake. They are also sites of attachment of soil mycorrhizae. Like all tip growing cells (pollen tube tip, fungal mycelia, mass protonema), root hairs grow by the addition of cell wall precursors to the tip delivered by vesicles from the endomembrane system. Because these non-dividing cells elongate outside the body of the root, their internal contents are easy to observe and experimentally manipulate. In recent years, root hairs are being used as experimental models to study biological processes.

Root Hair Initiation

Dolan *et al.* (1994) reported the details of root hair initiation in *Arabidopsis*. The root epidermal cells can be differentiated into two cell types, root hair cells (trichoblasts = hc) and hairless cells (atrichoblasts=nhc). In *Arabidopsis* epidermal cells are arranged in 16 to 22 cell files. Cells in 8 symmetrically positioned files differentiate into the hair cells, and the cells of the other files, into hairless cells. The position of root hair cell file is determined by the underlying cortical cells, which are eight in number. Root hairs always form from the epidermal cells positioned over the radial cell wall between cortical cells. Atrichoblast form from epidermal cells positioned on top of cortical cells. Root hair cells can also be cytologically distinguished by their dense cytoplasm than neighbouring epidermal cells at early stages. Thus, cell fate has been specified prior to cell elongation.

After specification, the epidermal cells that are committed to the root hair fate become wider, longer and deeper by diffuse growth. Before the hair begins to grow small GTP-binding proteins from the Rop family appear at the growth site. Within minutes of Rop protein localization the root hair cell wall begins to bulge out. Rop protein remains at the tip of the developing hair until growth ends. As the bulge enlarges, the endoplasmic reticulum within it condenses and actin (a component of microfilaments) accumulates. Under optimal conditions, it takes about 30 minutes for formation of root hair swelling on the surface of the cells.

Root Hair Tip Growth

Once a swelling is formed, the rest of the hair is developed by tip growth. During tip growth new cell wall material is synthesized and transported to the cell tip. The tip is packed with vesicles containing cell wall polysaccharides and cell wall proteins. These vesicles are produced by smooth and rough endoplasmic reticulum and dictyosomes. The nucleus enters the tube and moves along the tip until the tip is growing. Actin is concentrated at the tip throughout tip growth. Root hair tip growth requires calcium. The calcium gradient at the tip of the hair controls the direction of the growth. If the calcium ion concentration is artificially increased towards one side of the tip, the hair will re-orient to grow in that direction. Root proteins that are involved in root hair initiation also have a strong effect on the direction of tip growth. Microtubules also control the direction of the growth.

Cessation of Tip Growth

Root hair tip caeses its growth when hairs reach a predictable length. The end of growth is precisely controlled and produced a symmetrical, dome-shaped tip with the same diameter as the hair shaft. When hairs stop growing, the cytoplasm at the tip disperses and the vacuole enlarges into the dome. Rop protein, the calcium gradient and calcium channel activity are lost from the tip.

Regulators of Root Hair Growth

Phytohormones like kinetin, abscisic acid (at 10^{-9} to 10^{-5} M), brassinolide (10^{-10} to 10^{-6} M) and the synthetic hormone 2,4-D (10^{-14} to 10^{-5} M), stimulated root hair elongation(Izumo *et al.*, 1994). Marked promotion of growth was observed with 2,4-D (10^{-7} M) and kinetin (10^{-6} M). Further it was found that kinetin and abscisic acid caused branching and other morphological changes such as swelling of the tip or the base, twisting, curling and bending.

Dolan *et al.* (1994) have suggested that ethylene may induce trichoblast differentiation in the epidermis. Smith *et al.* (1994) have shown that the protein-phosphate inhibitors kadaic acid and calyculin-A (at concentrations as low as 3 nM) inhibits root hair growth and alter the cortical cell shape of *Arabidopsis* roots. The results indicate that these protein phosphatases have a role in controlling growth and development of root hairs.

Determination of Root Hair Development

In determining which epidermal cell of the root will become root hairs, dependent up on whether GL2 gene is up or down-regulated. If GL2 is upregulated the cell retains the atrichoblast fate. If GL2 is silenced, the cell becomes trichoblastic and a root hair develops. Here CAPRICE (CPC) protein promotes differentiation of hair-forming cells by controlling a negative regulator. GLABRA 2 (GL2) is preferentially expressed in hairless cells. It is observed that CPC is also predominantly expressed in the hairless cells, but not in neighbouring hair-forming cells. Then the CPC protein moves to the hair-forming cells through plasmodesmata and represses the GL2 expresseion. If the bHLH protein interacts with CPC protein, GL2 expression takes place. GL2 transcription stimulatory substance is thought to pass from the underlying cortical cells into the overlying epidermal cells. At the edges of cortical cells the apoplast between adjoining cells results in a region incapable of generating this message and transporting it into the overlying epidermis. Hence, epidermal cells positioned over a cell-cell junction in the cortical cells receive less of this stimulatory substance, resulting in the formation of trichoblast.

Galway *et al.* (1994) have shown that the TTG gene (transparent testa glabrous) of *Arabidopsis* is required to specify epidermal cell fate and cell patterning in the root, because the normal positioning of root hairs over the anticlinal walls of cortical cells is absent. They suggest that the TTG gene normally provides positional signals, so that differentiating epidermal cells to misinterpret their position and differentiate into inappropriate cell types. The ttg mutations were also associated with abnormalities in the morphology and organization of cells within and surrounding the root apical meristem and the gene may therefore have a fundamental role in development.

Functions of root hair:

(a) Root hair plays a role in water and mineral uptake.
(b) Root hair is an entry point for oxygen diffusion.
(c) Root hair is a site for mycorrhizal interaction.

Quiescent Centre

The middle tier of cells in the RAM comprises a central group of four large non-dividing cells known as the quiescent centre (QC). QC cells have a characteristic structure. QC maintains the initials in an undifferentiated state. This strategy ensures that after each initial division, one daughter cell is disconnected from the QC and allowed to differentiate. Clowes (1958) in his autoradiographic studies of DNA synthesis in the root tip of *Zea mays*, noted central cup-like region, the cells of which had fewer mitochondria and ER, very small nuclei and low rate of DNA and protein synthesis. He referred first this inactive region as 'quiescent centre'. Since then such regions have been reported in many plants. The cells of the quiescent centre do not divide but may become active when active initials are damaged. Thus, this region acts as reservoir of initial cell produicing cells. A ring of eight cortical endodermal initials that give rise to the cortex and endodermis surrounds quiescent centre.

Extensive research on root apex has shown that a group of cells at the pole of the stele and cortex have a very low mitotic activity. This group of cells, which appears in the shape of a hemisphere or a disc, was termed quiscent centre. The concept states that the most distal cells in the body of the root (the former initials of plerome and periblem) divide infrequently, show little change in size, and have low rates of synthesis of nucleic acids and protein. In apices with 125,000 to 250,000 meristematic cells the quiscent centre contains 500 to 1000 cells.

Various views have been expressed on possible causes of the appearance of a quiescent centre in roots. One view suggests control by hormones which may be synthesized in the quiescent centre, a substance may stimulate division at low concentration and inhibit at high concentration. Another view is that there is a competition between cells for supplies of nutrients or hormones. It was also suggested that the root cap controls Quiescence in the particular portion of the meristem. One of the regulator of growth in plants is pressure. Pressure exerted by the rapidly dividing neighbouring cells was thus suggested to cause the inactivity in the Quiescent Centre. The way the files of cells in the root tip run towards the Quiescent centre also suggests that the pressure extended by the expansion of the temporary initial (around the quiescent centre) and their derivatives prevents the quiescent centre from expending and thus its cells from dividing.

PRIMARY DIFFERENTIATION

At various distances from the apical meristem cells enlarge and develop their specific characteristics in relation to their position in the root; the cells become differentiated. The meristems of the three regions of the root, the epidermis, the cortex and the vascular cylinder, are delimited close to the apical meristem.

The protoderm of monocotyledon root is set apart as a derivative of the initials that also give rise to the cortex. In the dicotyledons, the protoderm is separated by a stepwise periclinal divisions (T divisions) from the derivatives of the layer of initials that forms the root cap. In roots with no distinct stories of intials, the protoderm has a common origin with the root cap and cortex and is not differentiated in early stages of development.

In the root region where vascular tissues are differentiating, the young epidermis proceeds with the formation of root hairs. Root hair-forming cells may not differ from those remaining devoid of hairs. In some plants, however, the protodermal cells that give rise to root hairs are smaller than their sister cells and show other features indicating specialization as hair-forming cells, or trichoblasts. Root hairs arise as small papillae in the region where cell division is subsiding and reach their full development at a root level where at least the first xylem is mature.

Cortical ground meristem consists of longitudinal files of cells attaining their maximum number by periclinal divisions (T divisions) close to the apex. The increase in circumference of the cortical layers occurs by anticlinal divisions. In many roots, the periclinal divisions progress in a centripetal order, that is they are repeated several times in the innermost layer of the nascent cortex. After the periclinal divisions in the inner cortex are completed the inner most layer differentiates as the endodermis.

The vascular cylinder is derived from procambium, or provascular tissue, appearing as a column in the centre of the root. The pericycle is usually the first cell layer in the vascular cylinder that can be identified as such. It delimits the vascular core close to the apical meristem. Within the core, enlargement and increasing vacuolation of certain cells block out the xylem pattern (diarch, triarch etc.) (Fig. 11.5). Because the metaxylem cells exceed the protoxylem cells in width, they are the first to become recognizable at the onset of enlargement and vacuolation. However, the deposition of secondary walls and functional maturation occur first in the protoxylem and later in the metaxylem. Thus, the primary xylem differentiates centripetally and is exarch. Phloem differentiation occurs in the same direction, that is, centripetally, for the protophloem appears next to the pericycle, the metaphloem deeper in the vascular core.

The longitudinal differentiation of the primary vascular tissues in the root is acropetal, the firat phloem maturing closer to the apical meristem than the first xylem (Fig. 11.5).

Lateral Roots

The lateral root primordia are formed from the pericycle cells just behind the root hair zone. Although all the cells in pericycle are capable of initiate lateral root growth, but only those that lie close to the xylem strands do this. The lateral root primordia usually arise in partially differentiated pericycle cells that have large vacuoles. But, by that time the adjacent endodermal cells have developed casparian strips. In corn, the cells of the pericycle and endodermis are thick-walled and lignified. The first visible change associated with the initiation of the primordium is a thinning of the wall and loss of lignin. The lateral root primordia do not arise from single cells in the pericycle but always form a group of cells. These are about 30 in *Raphanus* and 24 in *Vicia faba*. In *Arabidopsis*, approximately 14 pericycle cells become a lateral root primordium before the secondary root meristem appears. When they begin to divide, the cells become more densely cytoplasmic.

The pericyclic origin of the lateral root places it in close juxtaposition with the vascular tissues of the parent root with which the vascular tissues of the lateral root eventually become connected. The position of the lateral root with regard to xylem ridges of the parent root varies in relation to the vascular pattern of the parent root but is stable in a root with a given pattern. In a diarch root, the lateral root arises between phloem and xylem,

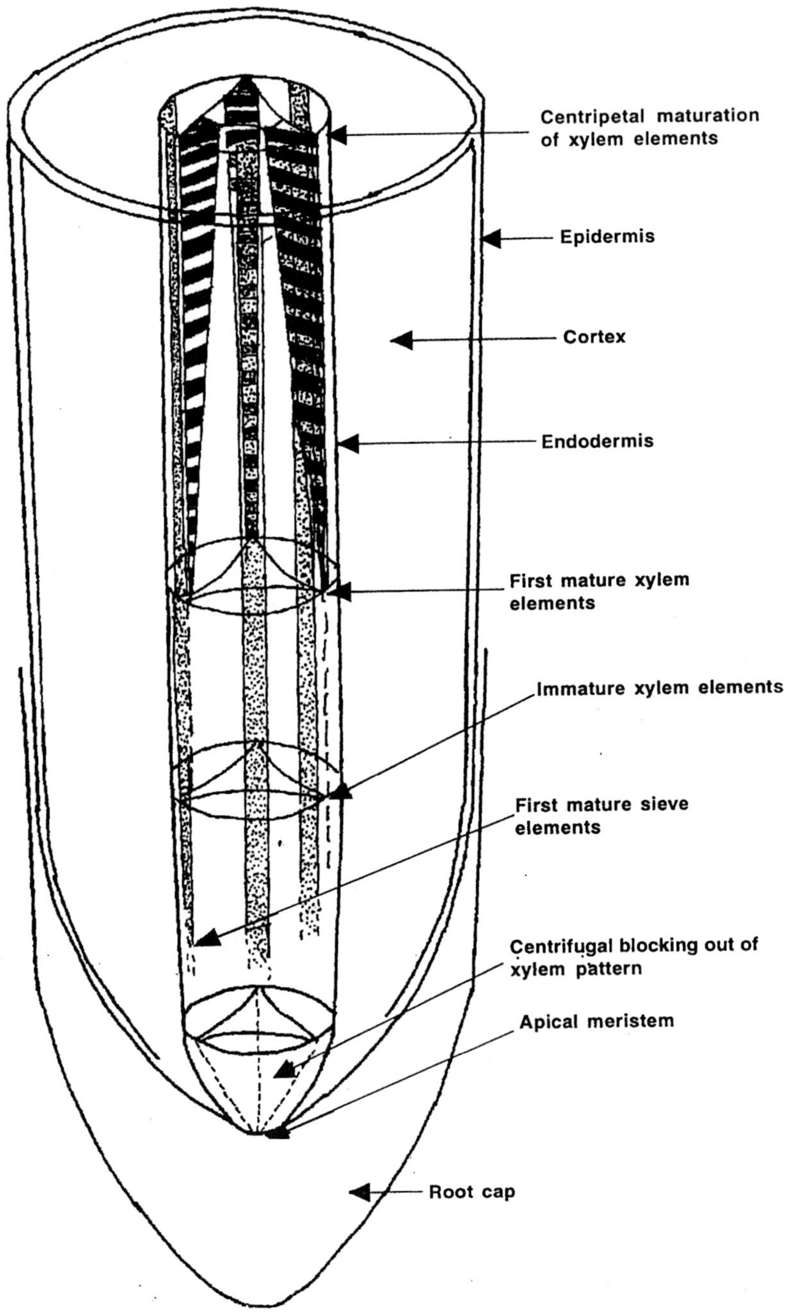

Figure 11.5: Diagram of primary vascular differentiation in a root of *Pisum sativum*.

in a triarch, tetrach and so forth, opposite the xylem, in a polyarch monocotyledon root, opposite the phloem.

When the phloem and xylem begin to differentiate in the lateral root these tissues become connected with the equivalent tissues in the parent root by a differentiation of the intervening parenchyma cells into vascular elements.

Lateral root initiation: One of the first events in root initiation is a periclinal division in a group of pericycle cells resulting in formation of two layers. They are designated as inner layer (IL) and outer layer (OL). Cells of the primordia begin to differentiate immediately after initiation.

Seedlings growth with limited nitrogen and high levels of sucrose initiate very few lateral roots. Sometimes, even the first division in the pericycle cells cannot be seen. Increasing nitrogen or decreasing sucrose restores initiation. High levels of salts such as NH_4NO_3 inhibit lateral root formation. Closer examination shows that lateral root initiation is unaffected, but lateral root primordia arrest take place. It appears that plants can regulate root system development in response to the environment both at the levels of initiation and at the level of development.

Lateral root development: Once the lateral root primordium is formed, it is differentiated into a mature lateral root by a two stage process. First, the primordium emerges through the overlaying tissues by cell expansion. The increase in cell size is apparent in cells near the base of the primordium, while cell number remains relatively unchanged. Secondly, the new lateral root begins to elongate, and cell numbers increase at the root tip, as the lateral root tip is meristematic. This is similar to that of the primary root elongation by division of cells in the root apical meristem. Due to periclinal divisions in outer and inner layer of cells two derivative cell layers are formed. They are designated as OL_1, OL_2, IL_1 and IL_2. OL_1 by one more periclinal division constitute the root cap and epidermis, OL_2 by periclinal division results in two layers of which the outer one constitute the cortex and inner layer develop into the endodermis. IL_1 develops into pericycle and IL_2 into vasculature. In all the angiosperms that have been studied, the endodermis contributes to the root primordium atleast in the initial stage. It usually divides anticlinalluy producing a unistratose sheath around the elongating primordium. Apparently, whatever tissue is provided by the endodermis is just temporary. It is ruptured even before the root emerges from the cortex. Exceptionally, it has been reported that in *Zea mays* the root cap of the new root is derived from the endodermis of the parent root. When the lateral root primordium has extended about 1/3 of the way through the cortex, the endodermal cells that cover its apex undergo periclinal division thereby producing a root cap meristem in *Zea mays*.

After completion of lateral root formation, it emerges out from the mother root. By that time, a well-defined apical meristem has been established and even a root cap may be present. The emergence of the root causes considerable damage to the cortex and epidermis and it results in an unprotected rupture that is a danger point for penetration by bacteria and fungi. A special corky tissue is formed and deposited around this to seal it off. As the lateral root begins to form its own endodermis, parenchyma cells at the point of origin of the root that connects the endodermis of the parent root.

This method of endogenous origin for root has extremely beneficial, because if the soil becomes dry the epidermis and cortex of the older portion of the root may die. But the endodermis, stele and other internal tissues remain alive. Lateral roots play an important role in water and nutrient uptake largely by increasing the root surface area.

Buds on Roots

Formation of buds on roots is a familiar phenomenon which is utilized in propagation of plants by root cuttings. Root buds frequently arise endogenously as do lateral roots. The earliest stages of root and bud differentiation appeared to be identical, but later stages differed not only in rates of development but also in histologic organization and kind of participation of parent tissues in the growth of primordium. Root growth is regulated by the gravity perception mechanism (statoliths) and a light perception (phytochrome), bud growth is regulated by light received by the root during its occasional approach to the surface of the soil. Buds on roots are seen in *Murraya koenigii* and *Millingtonia hortensis*.

Contractile Roots

Anchorage of a plant body by means of roots may involve root contraction, a process that pulls the shoot closer to the ground or in bulbous plants deeper into the soil. Root contraction is widely distributed among monocotyledones and herbaceous perennial dicotyledones. The contraction usually occurs in certain individual roots and is limited to parts of these roots. The shortening of the root depends mainly on the change in shape of the inner cortical cells which, at a certain stage of development expand radially and contract longitudinally. The root tissues not involved in this growth, that is central vascular tissues and the peripheral cortical tissues, become contorted and wrinkled.

Trichoblast: A cell in root epidermis that gives rise to a root hair is called trichoblast.

SECONDARY GROWTH IN DICOT ROOT

Most of the tap roots and main lateral roots of Dicotyledonous plants show secondary increase in thickness. As a rule, no secondary growth is present in

Monocotyledonous roots. The primary xylem (with exarch protoxylem) and phloem tissues are arranged in separate radii in the dicotyledonous roots. Pith is usually absent. Parenchyma cells present beneath each phloem patch become meristematic and form cambial strips. The number of cambial strips is equal to the number of phloem patches present. For example, a tetrarch root is with four strips of cambium and a triarch root with three strips. Subsequently the pericycle cells outside the xylem ridges also become meristematic and join the cambial strips formed beneath the phloem groups. The cambial ring thus is initially in the form of a wavy ring, and starts producing new derivatives by periclinal divisions both inwards and outwards. The cambium also increases in circumference by anticlinal divisions. This wavy cambial ring passes internal to the primary phloem and external to the primary xylem groups. Secondary xylem produced by the cambium is more in quantity than the secondary phloem. After a while, the wavy outline of the cambium assumes circular shape.

As a consequence of secondary xylem formation, the primary xylem is pushed to the central part of the root. The primary phloem elements especially the sieve elements get crushed. At this juncture, the root is identified mainly by the radial position of the exarch primary xylem in the central region. Normally ray initials are differentiated in the cambial ring in the region just opposite to the primary xylem. They ultimately produce broad vascular rays running across the secondary vascular tissues in roots. In addition to these, narrower phloem rays are seen within the phloem tissue.

To compensate the ruptured epiblem, due to increased diameter of the root consequent upon secondary thickening, periderm is produced in the outer tissues of the root. Basically the periderm of the roots is similar to that of the stem. The pericyclic cells undergo periclinal and anticlinal divisions. The periclinal divisions cause and increase of the number of pericyclic layers in radial extent. The combined increase in thickness of the vascular tissues and of the pericycle forces the cortex outward. The cortex does not undergo an increase in circumference but becomes ruptured and is cast of together with the epidermis and endodermis. A phellogen arises in the outer part of pericycle and forms phellem toward the outside. It may produce phelloderm toward the inside, but such phelloderm is difficult to distinguish from the pericycle which proliferated before the phellogen was initiated.

In perennial roots the activity of the vascular cambium continues through many years. The phellogen also continues its activity but may become replaced by phellogens arising at greater depths in the root. If such development occurs, the root, like a stem, has a rhytidome.

Anomalous secondary growth in roots is described elsewhere in the book.

CHAPTER XII

THE STEM

PRIMARY STRUCTURE

The stem consists of three tissue systems, the dermal, the fundamental and the fascicular or vascular.

Epidermis: Externally, the stem is bounded by the epidermis which contains epidermal cells, guard cells, idioblasts and different types of trichomes. The stem epidermis commonly consists of one layer of cells and has a cuticle and cutinized walls. It is a living tissue capable of mitotic activity. The epidermis responds to stresses by tangential enlargement and radial divisions.

Cortex and Pith: In Dicotyledonous stems, the ground tissue remains differentiated into hypodermis, cortex and pith. Such a differentiation is not found among monocotyledonous stem. Hypodermis in herbaceous stems is usually collenchymatous type which may be in the form of continuous ring as in *Aristolochia* or in the form of patches in angular stems such as *Cucurbita* and *Peristrophe.* Chloroplasts may or may not be present in these cells. In many stems, as for instance *Pelargonium*, *Retama* and *Salicornia* this parenchyma may have a photosynthetic function in addition to that of temporary storage of starch and other metabolites.

The cortex of stems contain parenchyma, usually with chloroplasts. Intercellular spaces are prominent but some times are largely restricted to the medium part of the cortex. In many aquatic angiosperms the cortex develops as an aerenchyma with a system of large intercellular spaces.

The stems of conifers and Angiosperms commonly lack a morphologically differentiated endodermis. In young stems, the innermost layer or layers may contain abundant starch and thus be recognized as a starch sheath. Some dicotyledons, however, do develop casparian strips in the innermost cortical layer of the stem and many lower vascular plants have a clearly differentiated stem endodermis.

The pith is commonly composed of parenchyma, which may contain chloroplasts. In many stems the central part of the pith is destroyed during growth. Frequently this destruction occurs only in the internodes, whereas the nodes retain their pith (nodal diaphragms). The pith has prominent intercellular spaces, at least in the central part. The peripheral part may be distinct from the inner having compactly arranged small cells and greater longevity. Since pith is also referred to as the medulla, the distinct peripheral zone of the pith is called perimedullary zone or medullary sheath.

Both cortex and pith may contain various idioblasts, including cells with crystals and other ergastic contents, and sclereides. If the plant has laticifers, such may be present in pith and cortex.

Vascular System: Internal to the cortex is the vascular system of the stem. In the Gymnosperms and most of the Dicotyledons the vascular system consists of a continuous or a split cylinder which encloses the pith. In this cylinder two types of vascular tissue can be distinguished, the phloem which is usually external and the xylem which is usually internal. In the case of the split cylinder each strand is termed a vascular bundle. A vascular bundle in which the phloem is only external to the xylem is said to be a collateral bundle. In some dicotyledonous families, e.g., Apocynaceae, Asclepiadaceae, Convolvulaceae, Cucurbitaceae, Solanaceae and certain tribes of Asteraceae, internal phloem is also present. This arrangement is called bicollateral vascular bundle. A stele in which the primary vascular tissue is arranged in strands around the pith is called Eustele, e.g., Dicotyledons, Gymnosperms.

The vascular bundles may be also concentric, in which either the phloem surrounds the xylem (amphicribal bundles) or the xylem surrounds the phloem (amphivasal bundles). The amphivasal bundles appear to be phylogenetically rather specialized and are found in certain positions in stems of some dicotyledons (medullary bundles in *Rheum, Rumex, Mesembryanthemum, Begonia*) and monocotyledons (Araceae, Liliaceae, Juncaceae, Cyperaceae). Amphicribal bundles are most common in ferns but are found also in angiosperms. Small bundles in flowers, fruits and ovules may be amphicribal.

Medullary and cortical bundles, present on the inside and outside of the stele respectively are found in certain plants. These bundles are associated with stems of both anomalous and typical structure. Medullary bundles occur in numerous dicotyledonous families, e.g., in the Amaranthaceae, Chenopodiaceae, Orobanchaceae, Berberidaceae and Cucurbitaceae. Cortical bundles are less common and are known to occur, for example, in the Melastomataceae, Proteaceae, Araliaceae and Calycanthaceae. So-called "cortical bundles" are often leaf-traces which descend through the cortex for some distance before entering the stele, e.g., as in *Begonia* and *Casuarina.*

In most monocotyledons and in a few dicotyledons no distinct vascular cylinder exists. The primary vascular system consists of a large number of bundles which are scattered irregularly and it is impossible to distinguish clearly the boundary between the cortex, the vascular cylinder and the pith. A stele with the vascular bundle as though scattered within the ground tissue is called atactostele.

Anatomy of the Node

In angiosperms the primary vascular cylinder is interrupted at each node by the exit of one or more bundles that enter the leaves. The stelar bundles which are the continuation of.bundles in the leaf bases, are called leaf traces. According to the number of leaf gaps per leaf the node is termed unilacunar (Fig. 12.1 A, D), trilacunar (Fig. 12.1 B, C) or multilacunar (Fig. 12.1 E). The traces also vary in length as measured by the number of internodes they traverse

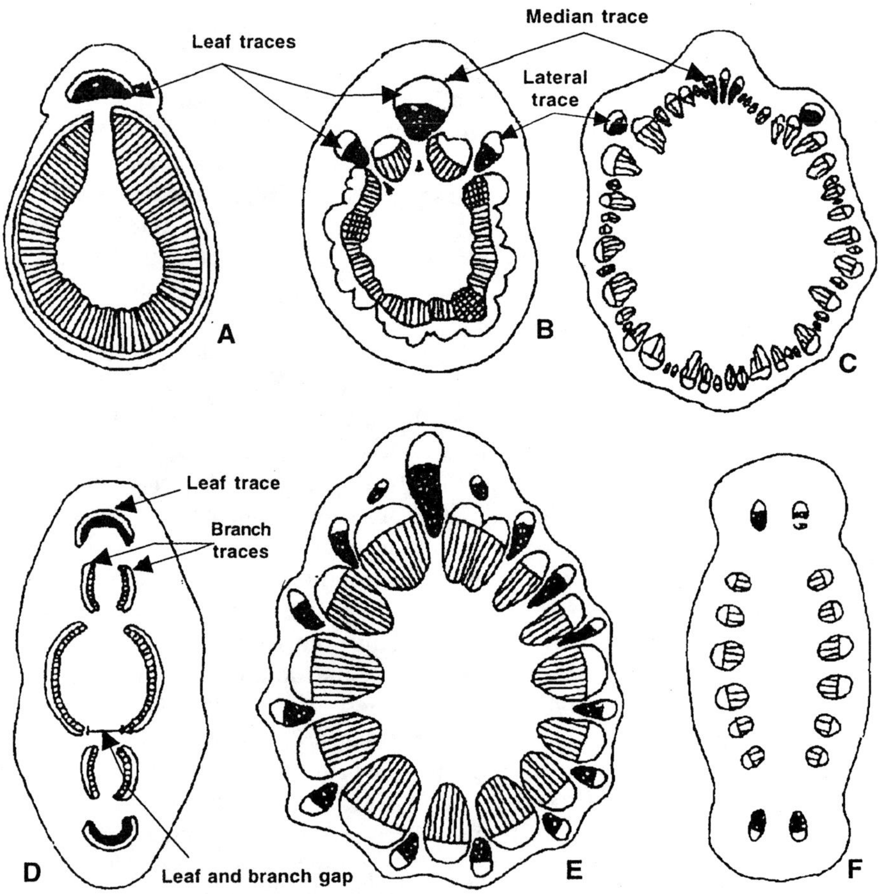

Figure 12.1: Cross sections of stems with different types of nodal structure. Leaf traces are indicated by blackened xylem regions. (A) *Spiraea*; (B) *Salix*; (C) *Brassica*; (D) *Veronica*; (E) *Rumex*; (F) *Clerodendron*.

before they diverge towards the leaf (less than 1 internode in the conifers, 1 to 35 in different species of dicotyledons.

The number of leaf traces and leaf gaps vary in different plants (Fig. 12.1) and may vary in the same plant at different levels. In most vascular plants only one trace is related to each gap. A deviating type of nodal structure is represented by the two-trace unilacunar condition (Fig. 12.1 F).

The anatomy of node is an important aspect of taxonomy. Sinnot (1914) concluded that the trilacunar node is the primitive type in the angiosperms and that the unilacunar node developed, phylogenetically, from it by the loss of two lateral gaps together with their respective traces or by approximation of the lateral traces to the median bundle to form a bundle composed of three traces, which is associated, therefore, with a single gap. The assumption that the trilacunar node is the primitive one among the Angiosperms has been refuted by later research on nodal anatomy in the gymnosperms and angiosperms (Bailey, 1956; Fahn and Bailey, 1957)

Branch Traces and Branch Gaps

Branches develop from axillary buds and have vascular connections with main axis. These vascular connections are termed branch traces (Fig. 12.2 B). At the node, the branch traces are situated very close to the leaf traces that enter the leaf in whose axil the branch develops. The branch gap is situated above the leaf gap and together these two gaps appear as one. In dicotyledons and gymnosperms the vascular supply to the axillary branches usually consists of two traces (Fig. 12.2 A). However, in some plants only one trace is present and in others more than two.

Leaf traces and stem bundles show structural differences. In transections of stems, the leaf traces are clearly circumscribed and have a larger amount of xylem than cauline bundles. As seen in longitudinal views, the leaf traces increase above their divergence from the stem bundles. Another notable feature of the leaf traces in members of diverse taxa is the occurrence of the vascular parenchyma cells differentiated as transfer cells with profusion of wall ingrowths.

Vascular System of Monocotyledons

The vascular system of monocotyledons usually consists of bundles that are scattered throughout the ground tissue of the stem. Among the dicotyledons, however, an arrangement of more or less scattered bundles does occur in a small number of plants, e.g., in the Nymphaeaceae, many plants of the Ranunculaceae and the herbaceous genera of the Berberidaceae.

In the Poaceae there are two basic types of arrangement of the vascular bundles: (1) that in which the vascular bundles are arranged in two circles,

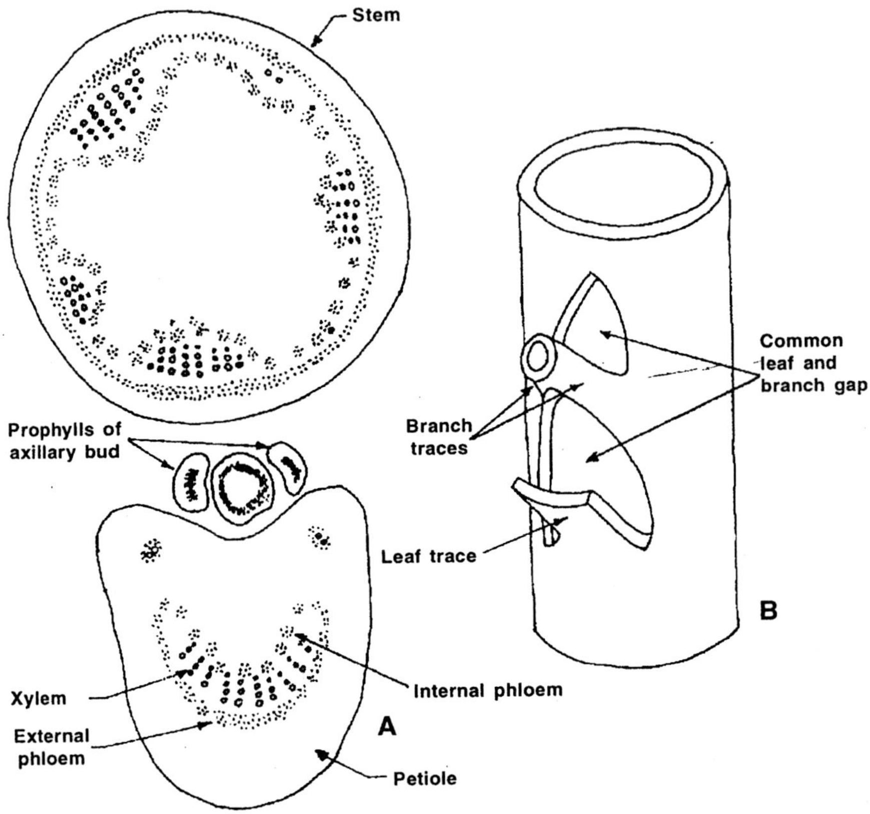

Figure 12.2: Relation between stem, leaf and axillary bud. (A) Transection of shoot of *Solanum tuberosum*; (B) Diagram showing relation of branch traces and a leaf trace to the vascular system in main stem.

the outer circle consisting of thin bundles and the inner thick bundles, e.g., *Triticum, Hordeum Avena* and *Oryza*; and (2) that in which the vascular bundles are scattered throughout the entire cross section of the stem, e.g. *Sorghum, Saccharum* and *Zea*. The individual bundles of the Gramineae are collateral and are usually surrounded by sclerenchymatous sheath.

TYPES OF STELE

The pattern of arrangement of the primary vascular tissues differs in the various plant groups and sometimes even within the same species. It has been found that the different patterns illustrate various stages in the evolution of the primary vascular system. The pattern in each plant is determined in the shoot and root apices from which the vascular tissues differentiate according to its genotype.

The stele of the sporophyte of vascular plants may be divided into two basic types: (1) *Protostele* which consists of a solid central cylinder of xylem surrounded by phloem; and (2) *siphonostele* in which there is a cylinder of pith within the xylem. Both ontogenetically and phylogenetically the protostele is the more primitive, and it is thought that the siphonostele has developed phylogenetically from the protostele.

Three types of protostele can be distinguished (1) haplostele which is the simplest type, in which the xylem appears more or less circular in cross-section, e.g., *Rhynia* and *Selaginella*; (2) actinostele in which the xylem is stellate in cross section, e.g., *Psilotum*; and (3) plectostele in which the xylem is split into longitudinal plates of which some are joined and others separate, e.g., *Lycopodium*.

The various types of protostele are characteristic of the Lycopsida (the lower Pteridophyta) and the siphonostele of the Pteropsida (the phylogenetically more advanced Pteridophyta and the Spermatophyta).

Two types of siphonostele are distinguished according to the positions of the phloem and xylem: (1) ectophloic siphonostele in which phloem only surrounds the xylem externally and (2) amphiphloic siphonostele in which the phloem surrounds the xylem both externally and internally and where the endodermis appears both outside and inside the vascular tissue on the borders of the cortex and pith respectively (e.g., *Adiantum* and *Marsilea*).

Siphonostele may consist of a continuous cylinder of vascular tissue or of a network of bundles. The latter type is the more advanced and the vascular tissues of this type appear in cross-section as a ring of separate bundles. The regions between the bundles are parenchymatous. Those parenchymatous regions that occur in the stele, above the positions where the leaf traces diverge from the stele to the leaves, are termed leaf gaps. An amphiploic siphonostele in which the successive leaf gaps are considerable distant, one from the other, is termed a *solenostele*. An amphiphloic siphonostele with overlapping gaps, i.e., that in which the lower part of one gap is parallel with the upper part of another gap, is termed *dictyostele*. In this case the bundles are interconnected to form a cylindrical network and each bundle is of concentric structure consisting of a central strand of xylem surrounded by phloem. Individually such bundles are termed meristeles. From the anatomical point of view these are amphicribal bundles.

During the course of evolutionary development the *eustele* with collateral bundles, developed from the ectophloic siphonostele. Bicollateral vascular bundles in which the xylem strands are accompanied externally by phloem strands and which are found in advanced Dicotyledonous families appear to be the result of a secondary specialization and not a relic of the primitive structure characteristic of the Filicinae. The vascular system of eustelic plants composed of bundles that branch and anastamose is referred to as a closed system. A vascular system characterized by bundles which branch, but still form independent, separate systems, is referred to as an open system.

In some plants, e.g., *Marattia*, *Pteridium*, and *Matonia*, two or more concentric cylinders of vascular tissue are present. Such a stele is termed a polycyclic stele. The individual cylinders in this case are interconnected. In rare cases stems and roots contain more than one stele; such condition is termed polystelic.

A different interpretation of the above nomenclature exists and has been summarized by Sporne (1962). According to this interpretation the ectophloic siphonostele without leafgaps, such as is found in the Pteridophyta, is considered as a protostele and is termed medullated protostele. Sporne does not use the term siphonostele for the Pteridophyta.

PRIMARY GROWTH IN STEMS

The axis of the embryo in the seed consists of a hypocotyl and radicle. At the tip of the hypocotyl one or more cotyledons and the bud of the shoot, i.e., the plumule, are found. At the tip of the radicle is the root cap. Both shoot and root apical meristems arise during embryogenesis and act throughout the lifetime of the plant as source of new cells and as sites of organ formation. It is at the shoot apical meristem (SAM) that leaf, stem and floral primordia are formed.

Angiosperm shoot apical meristem consist of a small number of morphologically undifferentiated, dividing cells located at the tips of shoots. The typical SAM is about 100 to 200 microns in diameter, although species with SAMs as small as 50 microns and as large as 900 microns have been reported. The morphology of SAM varies from flat (e.g., sunflower) through mound shaped (e.g., *Arabidopsis* vegetative SAM) to finger-shaped (e.g., maize) The size and shape of the SAM vary within a plant during the plastochron (the period between the formation of successive leaves). Toward the end of the plastochron, just before the emergence of a morphologically distinct primordium, the SAM is at its largest, whereas at the beginning of a plastochron, just after a leaf primordium has been produced, the SAM is at its smallest. Presumably, near the end of the plastochron the SAM includes a leaf primordium that is not yet morphologically distinct from SAM.

One of the most obvious features of the angiosperm SAM is its layered arrangement. The outermost region of the meristem consists of one or more sheets of cells that together comprise tunica. This sheet-like organisation is a consequence of the regular orientation of cell divisions within the tunica. The tunica overlies a mass of cells called corpus (Fig. 12.3 A, B). The irregular arrangement of cells within the corpus reflects the variable orientation of cell division in this region. However, in the basal/central region of

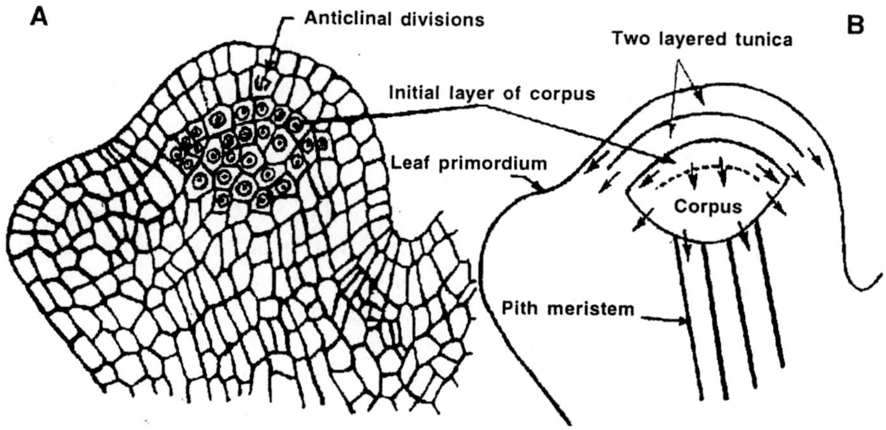

Figure 12.3: Shoot apex of *Pisum*. Cellular details in (A) interpretative diagram in (B) The pith meristem does not show typical rib-meristem form of growth.

the SAM, the rib meristem, composed of files of cells parallel to the long axis of the plant, can often be distinguished. Derivatives of the rib meristem give rise to internal layers of the stem.

Cell layers within the SAM are designated L1, L2, L3 etc., where L1 refers to the most superficial layer L2 to the next subjacent layer, and so on. Thus the *Arabidopsis* SAM consists of three layers L1 and L2 comprise tunica while L3 is the corpus. In maize, there are two layers in the SAM, the single-layered tunica (L1) and the corpus (L2). Despite the nearly universal occurrence of a stratified SAM development or function is unknown.

The bud of the shoot usually consists of an axis, the epicotyl, containing a few internodes, which have not elongated, and some leaf primordia. With the germination of the seed the embryo enlarges and starts to grow, the apical meristem of the young shoot adds further leaf primordia, and the internodes between the lower primordia which in the meantime have become distant from the apex, elongate. Buds develop in the axils of the developing leaves giving rise to a branched shoot. The details of shoot apex have been given earlier.

When the leaf primordium is initiated in the peripheral region of the shoot apex, localized cell divisions cause the formation of a protrusion (the so called leaf buttress) on the side of the axis. In shoots with a helical leaf arrangement, the divisions alternate in different sectors around the circumference of the apical meristem and the resulting periodic enlargement of the apex, as seen from the above, is asymmetric. In shoots with a decussate leaf arrangement the enlargement is symmetrical because the intensified meristematic activity occurs simultaneously on opposite sides. Thus, the initiation of leaves causes periodic changes in the size and form of the shoot apex.

The differentiation of procambium is illustrated in figure 12.4 in successive cross sections of a shoot of *Linum* (flax). No vascular differentiation is detectable above the level of leaf initiation (Fig. 12.4 A). Farther below where leaf primordia come into view, some vacuolation is present (Fig. 12.4 B). Increased vacuolation in cortex and pith, delimits an approximately circular zone as seen in transections (Fig. 12.4 C). For descriptive purposes, this meristematic zone is called residual meristem. At successively later stages of development more and more combial strands differentiate from the residual meristem (Fig. 12.4 D). After all vascular bundles characteristic of the given stem level have differentiated, the remaining residual meristem differentiates into interfascicular parenchyma (Fig. 12.4 E).

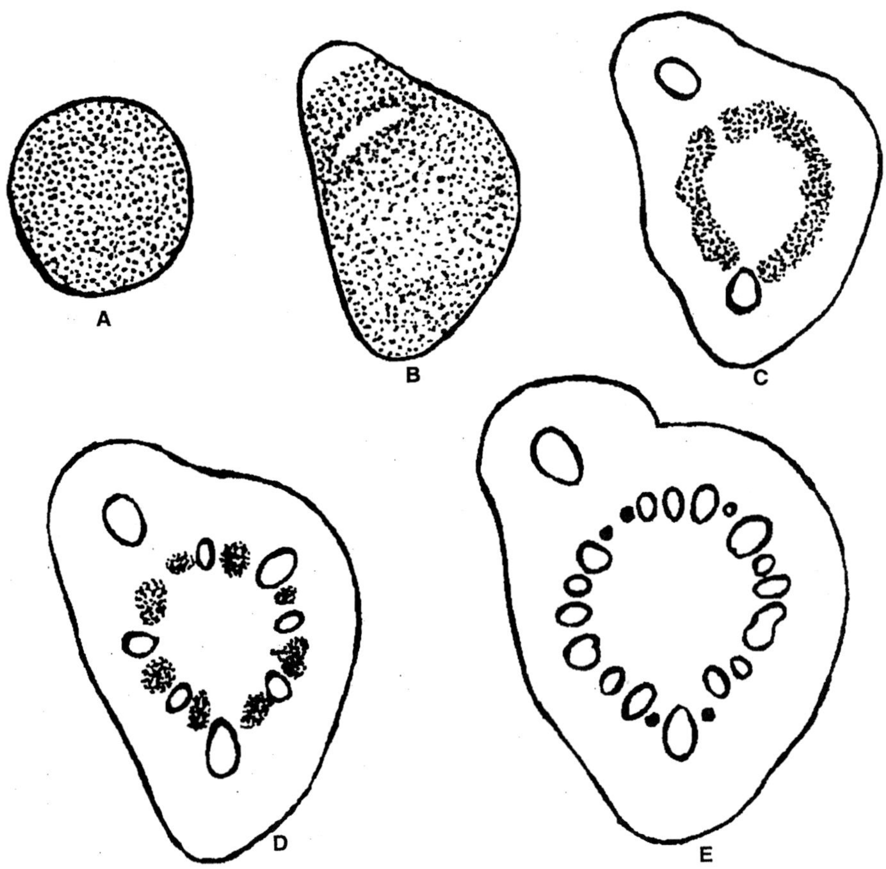

Figure 12.4: Cross sections of flax stem taken at different levels beginning with apical meristem (A) and ending with node in which all primary tissues are delimited; (E) Stages in differentiation of vascular system: vacuolation of cortex (starts in leaf primordium in B) and pith.

BRANCHING

There are two types of branching patterns in plants: terminal and lateral. Terminal branching involves the separation of the apical meristem into 2, which then both continue growing as a shoot branch. This mechanism is common in lower plants but rare in angiosperms. How the single integrated growing centre is divided is not understood. Experimental manipulations such as bisecting the meristem can artificially induce terminal branching, but these do not appear to mimic any naturally occurring events. They do speak to the capacity of isolated meristem segments to reorganize into complete meristems.

Lateral branching in seed plants occurs mainly by growth of axillary buds. When leaf primordia are initiated, a small group of cells in their axils remain small and unvacuolated. These are referred to as "detached meristems" because they retain meristematic characteristics and do not differentiate as the cells around them do, but they become spatially separated from the apical meristem. As the shoot grows and the detached meristem attain a distance from the apex, they organize into a shoot meristem with several leaf primordia (a bud). At this point their growth is arrested because of apical dominance (the shoot apical meristem inhibits growth of the axillary meristems by producing the hormone auxin). In bushy plants the arrest is temporary until the main shoot has grown beyond the inhibitory range of the apex. In unbranched plants the arrest is permanent unless the apical meristem is removed. Upon release from the apical inhibition, the lateral bud begins growth as a branch shoot.

SECONDARY GROWTH IN STEMS

Secondary growth in stem involves two processes — formation of secondary xylem and phloem and formation of periderm.

Formation of Secondary Xylem and Phloem

Secondary growth resulting from the activity of vascular cambium increase the amount of vascular tissues in stems. In dicot stems vascular bundles are collateral and conjoint and open with intra-fascicular or fascicular cambium. At the same time some cells between the vascular bundles lying in line with fascicular cambium also become meristematic and get differentiated as inter-fascicular cambium. Thus a complete ring of vascular cambium is formed.

The vascular cambium consists of two types of cells fusiform initials and ray initials. Fusiform initials are many times longer than wide, while ray initials are nearly isodiametric. Fusiform initials give rise to secondary xylem and phloem while ray initials give rise to vascular rays. When the cambial initials produce secondary xylem and phloem cells they divide periclinally. At one time a derivative cell is produced toward the xylem, at another time toward the phloem although not necessarily in alternation. During the height of cambial activity, cell addition occurs so rapidly that elder cells are still meristematic when new cells are produced by the initials. Thus a wide zone of more or less undifferentiated cells accumulates. Within this zone, the cambial zone, only one cell in a given radial file is considered to be the initial. The new cells formed towards the outer side facing the primary phloem get gradually modified into the elements of secondary phloem while those towards the inner side facing the primary xylem are modified into elements of secondary xylem. The ray initials of cambium cut off parenchymatous cells both outwards and inwards and thus form vascular rays.

Growth Layers — Annual Rings

In transverse section wood shows growth layers-growth rings. Each growth layer is divisible into early wood and latewood. The early wood, also known as spring wood, is less dense than the latewood, because wider cells with thinner walls predominate in the early wood, narrower cells with thicker walls in the late wood. The late wood forms a distinct boundary in a growth ring because of its sharp contrast to the early wood of the following season, but the change from early wood to late wood of the same growth layer is more or less gradual.

The relative amounts of early wood and latewood are affected by environmental conditions. Adverse growth conditions increase the relative amount of latewood in a pine but decrease the relative amount of late wood in an oak.

The successive annual rings formed year after year can be used in determining the age of a tree. Determination of the age of a tree by counting annual rings is called dendrochronology.

Distribution of Vessels (Ring porous wood and Diffuse porous wood)

The wood anatomist refers to a vessel in cross section as a pore. On the basis of distribution of pores in a growth layer two principal types of woods are recognised: diffuse porous wood with pores rather uniform in size and distribution throughout a growth ring, e.g., *Dalbergia, Shorea, Bombax, Artocarpus, Acacia, Pterocarpus, Terminalia, Mangifera, Fagus, Juglans, Liriodendron, Platanus, Populus, Pyrus* etc., ring porous wood with pores distinctly large in the early wood than in the latewood, e.g., *Castanea, Morus, Quercus, Ulmus, Robinia, Tectona, Lagerstroemia, Toona, Tilia* etc.

Distribution of Axial Parenchyma

The distribution of the axial xylem parenchyma shows many intergrading patterns. The spatial relation to vessels, as seen in transections serves for the division in two main patterns; apotracheal, parenchyma not definitely associated with the vessels (Fig. 12.5 A, B); paratracheal, parenchyma consistently associated with the vessels (Fig. 12.5 B, C). The apotracheal parenchyma is further subdivided into: diffuse, single parenchyma cells or parenchyma strands scattered among fibres; apotracheal banded (Fig. 12.5 B); boundary or marginal parenchyma, with single cells or a band at the end (terminal) or at the beginning (initial) of a growth layer (Fig. 12.5 A). Diffuse apotracheal parenchyma may be sparse. The paratracheal parenchyma appears in the following forms; scanty vasicentric, forming complete sheaths around vessels; aliform vasicentric with winglike tangential extensions (Fig. 12.5 C) and confluent, coalesced aliform forming irregular tangential or diagonal bands (Fig. 12.5 D). From the evolutionary aspect the apotracheal and diffuse patterns are primitive.

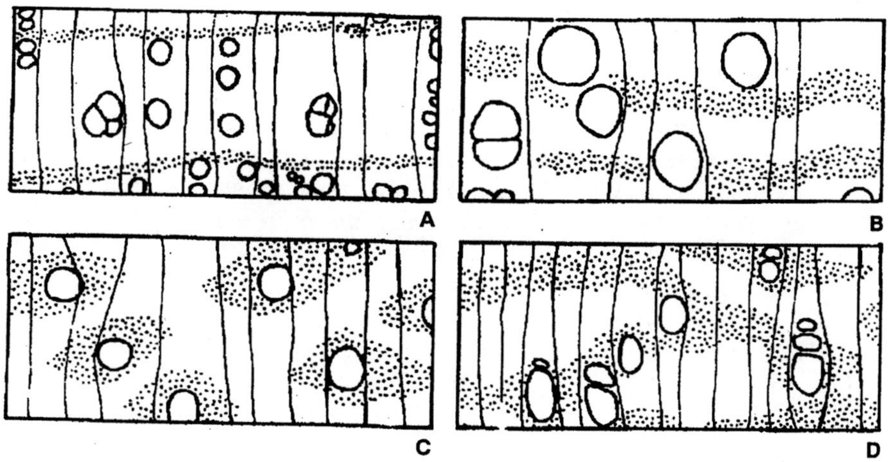

Figure 12.5: Distribution of axial parenchyma (stipples) in wood. (A) Boundary apotracheal — *Michelia*; (B) Banded apotracheal — *Saccopetalum*; (C) Vasicentric aliform paratracheal — a leguminous species; (D) Confluent paratracheal — *Terminalia*.

Heartwood and Sapwood

In a transverse section of a woody trunk, two regions, the outer light coloured and central dark coloured regions are seen. The light coloured region is called sapwood and dark coloured region is called heartwood. The sapwood, also known as albernum, consists of living and functional cells involved in conduction of water. Heartwood is also known as duramen.

The earlier increments of secondary xylem gradually become nonfunctional in conduction and storage and it is known as heartwood. Formation of heartwood involves removal of reserve materials or their conversion into heartwood substances and eventual death of protoplasts of the parenchymatic and other living cell types of the wood. Heartwood formation is explained as one of the processes enabling the plant to remove from the regions of growth those metabolic by products that may be inhibitory or even toxic to living cells. The translocation of these substances, possibly after their detoxification, occurs through the rays, in part towards the outer bark, in part towards the centre of the tree where the accumulation of excreta results in the death of cells. As the process continues the sapwood — heartwood boundary moves outward. The number of growth increments retaining sapwood characteristics varies greatly among different species.

In certain cases, distinction between sapwood and heartwood is not sharp as in *Holoptelia integrifolia* and *Abies pindrow*.

The pigments found in heartwood in certain cases are of commercial importance e.g., haematoxylin (*Haematoxylon campechiamum*), Brasilin (*Caesalpinia sappan*), Santalin (*Pterocarpus santalinus*).

Hardwood and Softwood

Woods are usually classified into two main groups, the softwoods and the hardwoods. The term softwood is applied to gymnosperm wood, that of hardwood to dicotyledon wood. The two kinds of wood show basic structural differences. The gymnosperm wood is homogeneous in structure — with only tracheids and fibre-tracheids. Dicotyledon wood contains vessels, tracheids, fibre-tracheids and libriform fibres.

On the basis of presence and absence of pores of commerce, i.e., vessels, wood is classified into porous and nonporous wood. The wood in which vessels are not found is called nonporous wood such as that of gymnosperms, whereas the dicotyledonous wood which has vessels in xylem is known as porous wood. The non-porous wood is technically called softwood and porous wood is called hardwood. Softwood or hardwood terms are not at all related with softness or hardness of timber, e.g., hardwood of *Bombax ceiba* (angiosperm) is very soft and softwood of *Pinus* and *Cedrus* (gymnosperms) is very hard. The wood of balsa (*Ochroma lagopus*), an angiosperm is the lightest and still classified as hardwood.

Reaction Wood

The reaction wood is formed on the lower sides of branches and leaning and crooked stems of conifer tree, and on the upper sides of similar structures in dicotyledonous trees. This wood is called compression wood in conifers and tension wood in dicotyledons. This wood is called reaction wood because its development is assumed to result from the tendency of the branch or stem to counteract the force inducing the inclined position.

The reaction wood shows structural differences. The compression wood is denser and darker than the surrounding tissue. Its trachieds are shorter, rounded with heavily lignified than those in the normal wood. In dicotyledons the vessels of tension wood are reduced in number and width. The wood has also got characteristic presence of gelatinous fibres which have a thick highly refractive gelatinous layer in the cell wall.

Storied Wood and Nonstoried Wood

In transverse sections the secondary xylem shows more or less orderly radial seriation of cells — a result of the origin of cells from tangentially dividing cambial cells. In the homogenous conifer wood this seriation is pronounced; in vessel-containing dicotyledons it may be somewhat obscured by the ontogenetic enlargement of the vessel members and the consequent displacement of adjacent cells. Radial sections also reveal the radial seriation, and they indicate that the radial series of the axial system are superimposed one upon the other in horizontal layers, or tiers. The tangential sections are varied in their appearance in different woods. In some, the cells of one tier unevenly overlap those of another, in others the horizontal layers are clearly displayed in tangential sections. Thus, some woods are nonstratified or nonstoired, in tangential sections (*Castanea, Fraxinus, Juglans, Quercus*), others stratified or storied (*Aesculus, Cryptocarya, Ficus, Tilia,* numerous Fabales). The storied condition is especially pronounced when height of the ray matches that of a horizontal layer of the axial system. In storied xylem the vessel members are usually short. From the evolutionary aspect the storied woods are more highly specialized than the nonstoried.

Vascular Rays

A panel of tissue variable in height and width, formed by the ray initials in the vascular cambium and extending radially in secondary xylem and secondary phloem is known as vascular ray.

Vascular rays may be uniseriate (one celled in width) or multiseriate (two or more celled in width).

The radial parenchyma cells are divided into categories according to their form. The two most common types are the procumbent and the upright ray cells (Fig. 6.3). A ray cell the longest diameter of which is oriented radially is procumbent; the cell elongated axially is upright. The two kinds of ray cells are often combined in the same ray, the upright cells appearing at the upper and lower margins of the ray. Rays composed of one kind of cell are called homocellular (Fig. 6.3 C, D), those containing procumbent and upright cells heterocellular (Fig. 6.3 A, B).

The rays of confiers are mostly one cell wide, occasionally biseriate and from one to twenty or even to fifty cells high. Presence of resin ducts makes the normally uniseriate appear multiseriate.

In contrast to the predominantly uniseriate rays of conifers, those of dicotyledons may be one to many cells wide, that is, they may be uniseriate or multiseriate and range in height from one to many cells. The multiseriate rays frequently have uniseriate margins.

The parenchyma cells of vascular rays are involved in storage and translocation of oil and many other ergastic substances. Tanniferous compounds and crystals are common inclusions in vascular rays.

Tyloses

In many species parenchyma cells (axial or one in a ray) located next to the vessels form outgrowths through the pit cavities into the lumina of the vessels (Fig. 12.6) when the latter become inactive. Soon the wall of the parenchyma also gets thickened due to gradual deposition of lignin. The nucleus and part of the cytoplasm of the parechyma cell commonly migrate into the tylose.

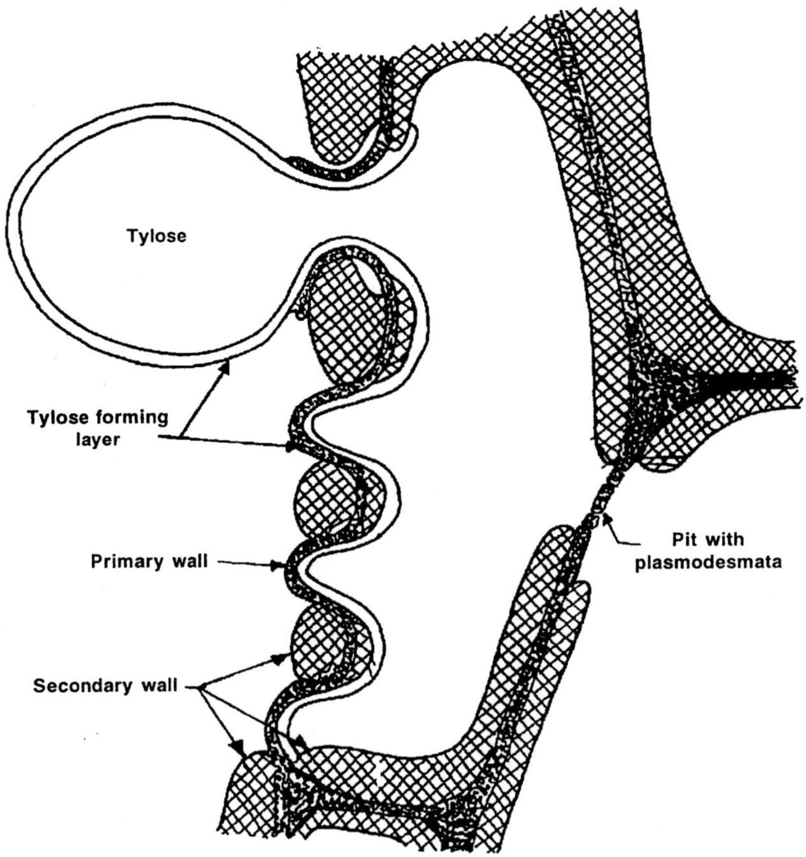

Figure 12.6: Diagram of a ray cell that has formed a tylose.

These outgrowths are called tyloses. Tyloses block lumina of vessels and reduce the permeability of the wood. Tyloses block vessels during the formation of heartwood and in the sapwood, in response to wounding and infection and diseases. Tyloses development is possible only if the pit aperture on the vessel side is no less than 10 microns wide. Examples of woods with abundant development of tyloses are *Quercus, Robinia, Vitis, Juglans* and *Morus*. Tyloses occur in the primary xylem also and are seen in coleoptile and bean leaves. Tyloses also occur in certain herbaceous plants like *Cucurbita, Coleus* and *Canna*.

Formation of tyloses is important in the treatment of wood with preservatives and in its selection for tight coopering. An outgrowth resembling a tylose is known as tylosoid. Parenchyma cells show outgrowths into sieve elements (phloem) and these are known as tylosoids. Similarly epithelial cells show out growths into intercellular resin ducts.

Resin Ducts

In most of the gymnosperms, resin ducts are developed in the axial or ray systems. These ducts develop schiziogenously between resin producing parenchyma cells, which then form the epithelium of the duct. Sometimes the resin duct may become blocked by enlargement of the epithelial cells and these structures are called as the tylosoids. Differences exist in the thickness and lignification of cell wall of the epithelial cells in various coniferous genera. In those genera, where the resin secreting cells have thick, lignified walls and in which these cells die after one season, relatively little resin is produced. In those plants where the epithelial cells are thin walled and function during several seasons, a large amount of resin is produced. Resin ducts with lignified epithelial cell walls are common in *Abies* and *Cedrus*, while thin walled, non-lignified secretory cells occur in *Pinus*.

Resin ducts appear as a normal feature of the wood in *Pinus, Picea, Larix* and *Psudotsuga*. However, these are also produced as a result of injuries, such as wounding, pressure and frost. In certain conifers like *Cupressus*, resin ducts are never developed in the secondary xylem.

Gum Ducts

Intercellular gum ducts occur in dicot wood similar to the resin ducts of the gymnosperms. They may arise normally in the wood or may be the result of various kinds of injury, e.g., Simaroubaceae. These ducts contain various substances such as resins, oils, gums and mucilages. Normally formed gum ducts always occupy a characteristic position in the wood, while those of the traumatic origin may be variously situated. In a given plant the canals may be either horizontal or vertical, but only in a few cases are the two types present in a single plant.

CHAPTER XIII

ANOMALOUS SECONDARY GROWTH

The dicotyledons and gymnosperms exhibit continued accretion of secondary vascular tissues all through their life by the activity of the normal vascular cambium. This is usually described as normal secondary growth. In a number of angiosperms, however, secondary growth takes place in a way that is different from that of normal type, and is then described as anomalous secondary growth. This is also termed abnormal secondary growth or unusual secondary growth.

According to Haberlandt (1914), from the physilogical point of view, the anomalous secondary growth may be of the Non-adaptive and Adaptive types.

Non-Adaptive type: The anomalous secondary growth exhibited by some herbaceous plants (e.g., *Amaranthus, Achyranthes, Boerhavia*) is not of any functional significance to these plants and may be described as of non-adaptive in nature.

Adaptive type: The anomalous secondary growth is of definite functional advantage to the plant e.g., *Bignonia, Bauhinia, Serjania*. The mechanical requirements of stem of woody climbers are quite different from those of the ordinary woody plants. The stem of a woody climber must be flexible so that it would be able to twist and twine around its support in an efficient way and at the same time it should be strong enough to support the weight of its own foliage. The above factors are responsible for the anomalous secondary growth in these plants. Further, a solid mass of vertical trunk does not serve the purpose of a woody climber. On the contrary, a woody climber requires a stem that is supple and flexible enough to enable it to survive around its support efficiently and with ease. As a result of anomalous secondary growth, such a type of stem is produced in a woody climber.

METHODS OF ABNORMAL GROWTH

Anomolous Secondary growth takes place through one of the following methods.

1. Abnormal activity of the cambium
2. Abnormal position of the cambium
3. Formation of more than one cambial ring
4. Formation of inter xylary (included) phloem
5. Formation of inter xylary cork.

Abnormal Activity of the Cambium

Bauhinia: In *Bauhinia rubiginosa* the cambium is more active at certain points but shows limited activity or no activity at other regions. In the regions of more activity, excessive secondary vascular tissues is produced, where as in the regions of less activity, little or no vascular tissue is formed (Fig. 13.1 D). Such an abnormal growth pattern results in the formation of ridged vascular cylinder and the mature stem exbibits a ridged appearance externally. In other species of *Bauhinia*, the activity of the cambium is limited to the two opposite poles only, resulting in an elongated flattened strap-like structure (Fig. 13.1 F).

Bignonia: In *Bignonia* the output of secondary xylem is relatively less from the beginning at four points, which are arranged in the form of a cross. At these four points however the secondary phloem tissues are more. Consequently, the stem appears to be provided with four longitudinal furrows. These grooves become constantly deeper, as the secondary growth proceeds. Later, the woody cylinders become split up into seperate strips.

Representatives of Bignoniaceae illustrate yet another type of anomalous growth. After the usual type of cambial cylinder is formed at the end of primary growth, four strips of cambium caese to produce xylem but continue to give off derivatives to the phloem side. Thus, two kinds of cambium become established, one showing bidirectional activity, the other unidirectional activity. As the stem increases in circumference additional strips of cambium change from bidirectional to unidirectional cell production. Some of the new phloem panels develop next to the earlier ones, others appear in new regions of the xylem. Each phloem panel is flanked by wide rays and retains its original tangential diameter, but the groups of panels increase in width towards the outside through the addition of new panels to each group (Fig. 13.1 A). The xylem is thus repeatedly fissured by the insertion of new phloem panels. In old stems, it is further broken up by dilation growth of xylem and pith parenchyma.

Where a cambium, normal in position and activity forms in some segments much larger proportions of xylem than of phloem, and in others more

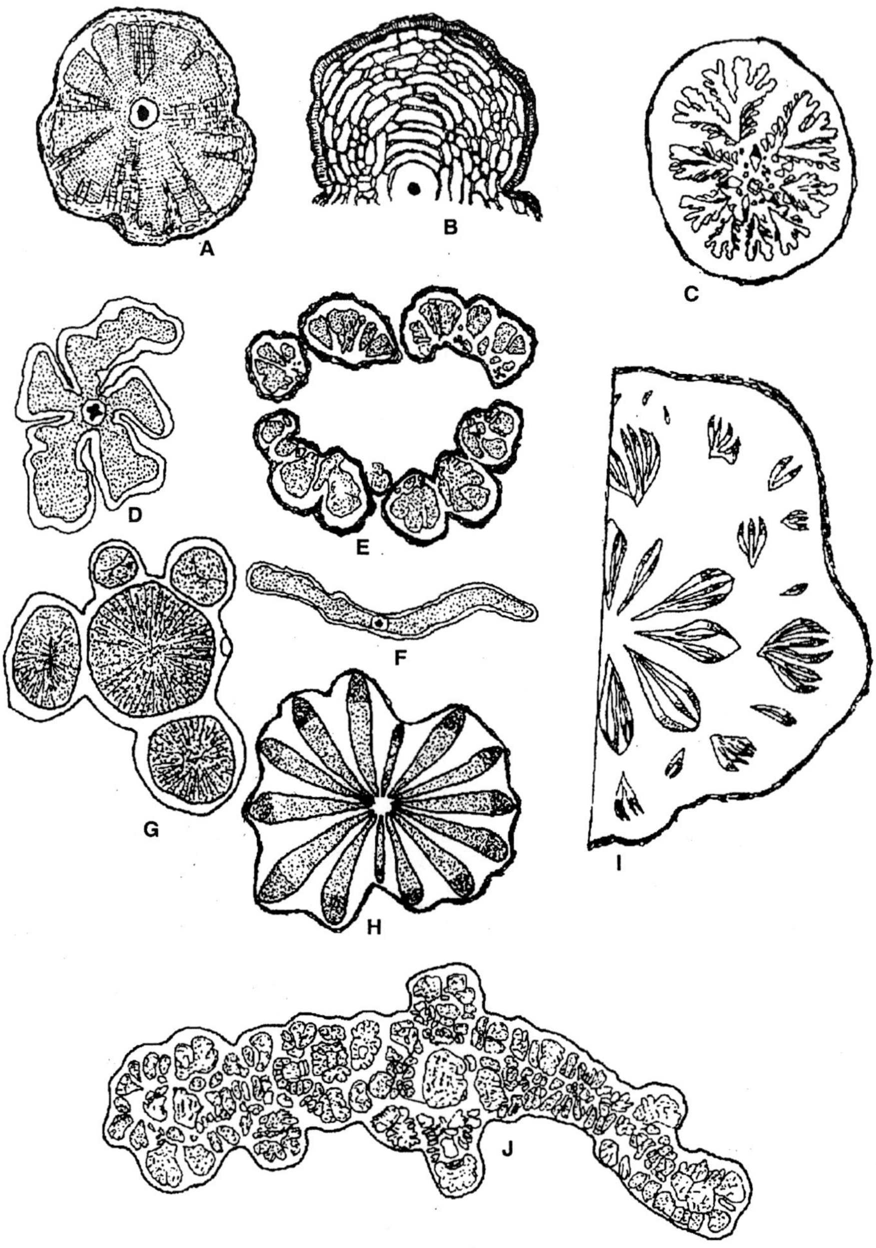

Figure 13.1: Anomalous stem structure — liana types. (A) A Bignoniaceous genus; (B) *Securidaca lanceolata*; (C) *Bignonia* sp.; (D) *Bauhinia rubiginosa*; (E) *Serjania ichthyoctona*; (F) *Bauhinia* sp.; (G) *Thionouia scandens*; (H) *Begonia fruticosa*; (I) *Piper fluminence*; (J) *Bauhinia langsdorfiana*.

phloem than xylem, a ridged and furrowed xylem cylinder is formed. This may be simple in structure as in Fig. 13.1 A or very complex as in Fig, 13.1 C. In a cross section of a stem, the cambium produces more xylem than phloem in certain places and more phloem than xylem in others, the xylem cylinder becomes ridged and sometimes even more complex structures may develop. Such a stem configuration occurs for instance, in *Passiflora glandulosa* (Fig. 13.2 C).

Urtica dioica stem: In young stem a normal cambial ring is formed which behaves normally cutting off secondary phloem towards outside and secondary xylem towards inside. However, after sometime the cambium begins to behave abnormally at certain places only cutting off parenchymatous cells on the inner side, instead of secondary xylem. After cutting off a group of parenchymatous cells the cambium again begins to behave normally and resumes its normal activity, forming secondary xylem above the parenchymatous group. The process is repeated again and again, resulting in the formation of islands of parenchyma which are embedded in the secondary xylem.

The cambium forms vascular tissues only in the region of vascular bundle:

Aristolochia: During the secondary growth, the parenchymatous cells of the medullary rays adjacent to the fascicular cambium become meristemataic to form interfascicular strips of cambium. The interfascicular cambium behave abnormally by producing only the parenchyma cells. The fascicular cambium, on the other hand, behaves normally producing secondary xylem and phloem within the vascular bundles. Concomitantly, the vascular bundles, gradually increase in size but continue to maintain their individuality (Fig. 13.2 A). Such a condition is also seen in *Begonia fruticosa* (Fig. 13.1 H).

Cucurbita: In *Cucurbita* also the fascicular cambium produces xylem towards inside and phloem towards outside while interfascicular cambium produces only parenchyma cells.

Abnormal Position of the Cambium

(a) Thionouia scandens (woody climber of Bignoniaceae)

In young stems the cambium is not circular but is with ridges. Later, the cambial ridges get seperated and act as individual cambia forming individual groups of vascular tissues. Consequently the secondary growth becomes abnormal with separate steles with individual secondary phloem and secondary xylem resulting in a peculiar shape in the mature stem (Fig. 13.1 G).

(b) Serjania and Paulinia (Sapindaceae)

In these tendril climbers, specific groups of vascular tissue are constricted off from the main vascular cylinder. These constricted off vascular tissues

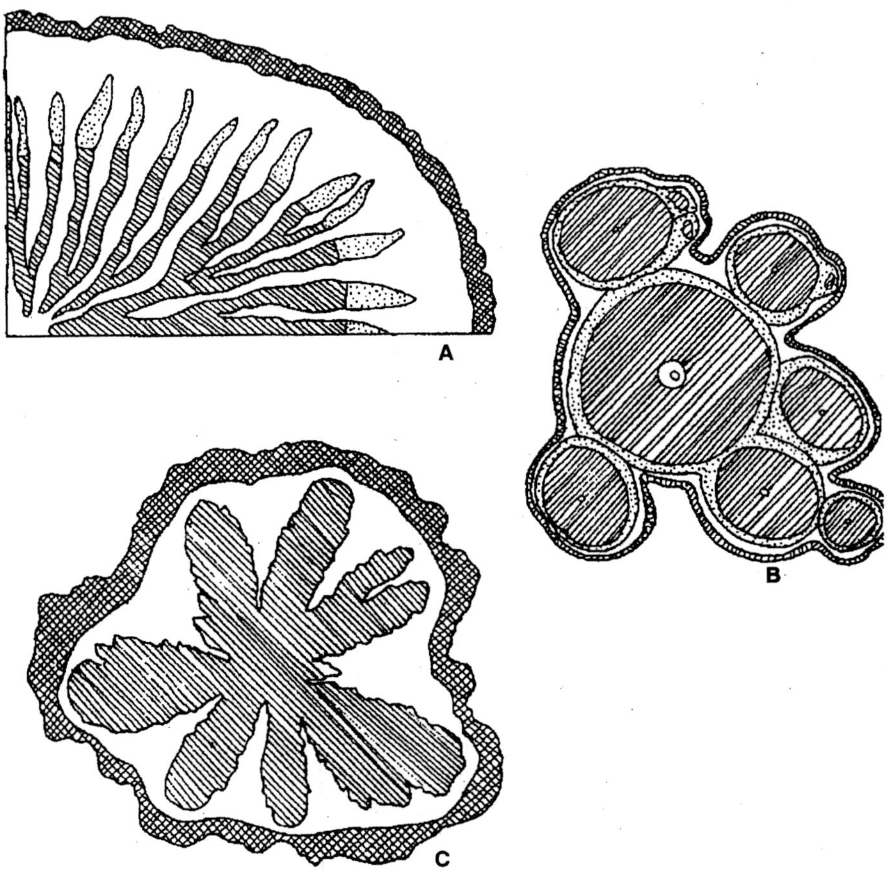

Figure 13.2: Anomalous stem structure. (A) *Aristolochia triangularis*; (B) *Serjania clematidifolia*; (C) *Passiflora* sp.

develop their individual cambial strips. With the initiation of secondary growth, each strip of cambium develops to form a complete and separate cambium ring, which function normally forming secondary xylem on the inner side and secondary phloem on the outer side. Thus, the mature stem consists of a number of discrete vascular cylinders each of which develops its own periderm. In the mature condition the stem seems to be made up of several smaller stems appressed to each other and giving the appearance of strands of rope (Fig. 13.2 B). This apparently compound condition becomes more marked as the stem becomes older and the parts separate as the outer laters of each strand die because of the development of periderm layers (Fig. 13.1 E).

(c) Bauhinia landsdorffiana (Fam: Fabaceae)

In this plant the parenchyma of secondary xylem and secondary phloem tissues undergo intercalary growth resulting in excessive amounts of parenchymatous tissues termed dilation parenchyma. This makes the secondary tissues along with cambium split into many strands (Fig. 13.1 J).

Abnormal position of the cambium is also seen in securidaca (Fig. 13.1 B).

Formation of More than One Cambial Ring

In some angiosperms (*Boerhavia diffusa, Bougainvillea, Mirabilis*), normal vascular cambium caeses to function after the production of a limited quantity of vascular tissues. New cambia are, however, produced successively, in the pericycle and cortex. About 28 families of Dicotyledons show formation of accessory cambia (Fig. 13.1 I). The cambium in this zone behaves abnormally producing vascular tissue in the fascicular region and lignified conjunctive tissue in the interfascicular regions.

Successive rings of cambia are formed which behave in a manner similar to the first ring, as a result, concentric layers of vascular bundles are formed embedded in conjunctive tissue.

Formation of Inter Xylary Phloem

Variations of another type in the activity of the cambium produce inter xylary phloem. Phloem of this type is secondary phloem in the form of strands embedded in secondary xylem. There appears to be two methods by which inter xylary phloem comes to lie embedded in secondary Xylem.

Interxylary phloem is seen in about 25 families of Dicotyledons like Combretaceae, Malpighiaceae, Solvadoraceae, Asclepiadaceae, Acanthaceae, Gentianaceae etc.

(a) In genera like *Combretum* and *Entada*, small segments of the cambium are said to produce phloem cells toward the inside for a brief period, in place of the xylem cells which are normally produced. After a period of such activity these cambium segments return to normal function, and thus bury the inwardly formed phloem with xylem.

(b) In other forms, such as *Strychnos*, the interxylary phloem strands are formed by the cambium toward the outside as a part of the normal external phloem, but the strands later become embedded in the xylem in the following manner. Small segments of the cambium cease to function, their cells becoming transformed into mature conducting tissue. New segments of cambium then arise, as secondary meristems, in the phloem some few rows of cells out from the original camium or in the pericycle. These unite with the edges of the segments of the general cambium cylinder — which have continued meanwhile their normal activity — and thus enclose a strand of phloem cells. This is repeated in other segments of the cambium so that the secondary xylem possesses numerous scattered strands of embedded phloem.

(c) In *Leptadenia* (Asclepiadaceae) and *Thunbergia* (Acanthaceae) also phloem is formed not only toward the outside but from time to time also toward the inside.

Interxylary phloem plays siginificant role in some xerophytic plants as it remains functionally active even during drought conditions when other tissues desiccate and become non-functional.

(d) *Salvadora persica*: The Cambium functions normally cutting off secondary xylem on inner side and secondary phloem on outer side. After sometime at certain points, the cambium begins to behave abnormally cutting parenchymal cells instead of secondary xylem on the inner side. This is, however, a temporary change and soon the abnormally behaving cambium strip reverts back to its original function of cutting secondary xylem on the inner side. This abnormal behaviour is repeated several times, resulting in the formation of a number of patches islands of thin walled cells. The parenchymatous cells ultimately get converted to phloem parenchyma (peripheral cells) and sieve tubes and companion cells (central cells) giving rise to phloem patches termed as included phloem or inter xylary phloem.

Formation of Interxylary Cork

In several plants such as *Achillea fragmentissima*, *Artemisia tridentata* and *Epilobium angustifolium* interxylary cork is formed. The cambium ring cuts off secondary xylem and parenchyma towards the inner side. Soon a layer of phellogen (cork cambium) develops in this parenchymatous zone between the secondary xylem. This layer of phellogen becomes active and cuts off cork cells which remain enclosed between two consecutive rings of secondary xylem, and is termed interxylary cork. In older stages, suberization of the medullary rays may occur or certain portions of the cambium may cease its activity resulting in splitting of the stem.

Anomalous Secondary Growth in Monocot Stems

In Monocotyledonous plants, secondary growth is usually absent as the vascular bundles are of closed type, i.e., cambium is absent. But some members of Liliflorae like *Dracaena, Yucca, Aloe, Agave, Cordyline, Sansevieria, Lomandra, Xanthorrhoea* and *Kingia* exhibit secondary growth which is anomalous in origin as well as in function. The cambium arises in the parenchyma of cortex outside the primary vascular bundles. The activity of this cambium is unusual. It produces more parenchyma towards the inside and a small amount of parenchyma toward the outside. In the development of the vascular bundles, individuals cells derived from the cambium divide longtudinally, then two or three of the resulting cells repeat the longitudinal divisions. The products of the final divisions differentiate into vascular elements and associated sclerenchyma cells. Vertically, many tiers of cells combine to form a vascular bundle. The secondary vascular bundles may be collateral or amphivasal and occur in more or less definite radial files. The cells formed outwards by cambium are parenchymatous and constitute secondary cortex. They are thick-walled, lignified and are arranged in regular radial rows.

Formation of storied cork: The parenchyma of the cortex divide many times periclinally and the products of these divisions, aligned in distinct radial rows (storied cambium) become suberized. The cork of this type is called the storied cork.

ANOMALOUS SECONDARY GROWTH IN ROOT

In a number of dicot plants, the roots are modified to function as storage roots. In the roots of Apiaceae as those of Fennel (*Foeniculum*) or carrot (*Daucus*) the secondary growth is of the ordinary type but parenchyma predominates in the xylem and phloem. While in other like *Beta* and *Ipomoea batatas* secondary growth in roots is usually of anomalous type as an adaptation to their storage function. Members of Apiaceae, Brassicaceae, Chenopodiaceae, Asteraceae, Convolvulaceae etc. show anomalous secondary growth in roots.

(a) *Beta vulgaris*: The secondary growth is initiated by the formation of a cambium in the interstitial parenchyma, except opposite the two protoxylem groups where it is from the pericycle. The cambium gives rise to a ring of essentially collateral vascular bundles seperated by narrow bands of parenchyma. After some time, the cambium ceases to function and another cambium arises from the phloem parenchyma of the first ring of vascular bundles. This cambium produces a second ring of vascular bundles separated by large radial channels of storage parenchyma. Soon a third ring of vascular bundles is produced from yet another cambium derived either from phloem or pericycle, and it also behaves in the same way as the earlier cambia. Thus, a number of accessary cambia are produced and contribute to the successive rings of vascular bundles alternating with parenchyma. The vascular bundles produced in successive rings by the accessary cambia consist more of parenchyma and less of xylem elements.

(b) *Ipomoea batatas*: Another complex type of anomalous growth occurs in the fleshy adventitious roots of *Ipomoea batatas* (Sweet Potato). Secondary growth of the storage root takes place by the formation of cambium ring which is normal in function intially. The xylem contains a large proportion of parenchyma. Cambia develop in the parenchyma around individual vessels or vessel groups and produce a few tracheary elements toward the vessels, a few sieve tubes and laticifers away from the vessels and a considerable number of storage parenchyma cells in both directions. Thus, phloem elements appear within the part of the root that originally differentiatied as xylem.

CHAPTER XIV

THE LEAF — STRUCTURE AND DEVELOPMENT

The leaf is one of the three basic organs of higher plants along with the stem and root. It plays a vital role in Photosynthesis. The leaf is the most variable of all the plant organs. 'Phyllome' is the collective term generally used for the various types of leaves appearing in plants. The phyllome of the spermatophyta is quite variable in its external and internal structure as well as in its function. The phyllomes are thus classified as: Foliage leaves, Cataphylls, Hypsophylls, Cotyledons etc. The foliage leaves are the main photosynthetic organs. The cataphylls are the scales which appear on buds and underground stems and function as protective agents as well as storage organs. The hypsophylls are the bracts present in the flowers and they also help flowers in protection. Rarely, they are coloured and function like petals. The 'Cotyledons' are the first leaves of the plant. The floral organs are also considered as the leaves.

The leaves are divided into simple and compound basing on whether the stalk bears one or more leaflets. The common stalk of the compound leaves is known as the rachis. If the leaflets arise from the sides of the rachis, the leaf is called pinnate and if all of them arise from one point, as rays, the leaf is palmate. Mostly, the leaves are dorsiventral and flattened laterally, as in *Iris* or cylindrical as in *Sansevieria* or may be tubular in some insectivorous plant. In certain plants like *Acacia*, the lamina of the leaves is reduced and the remaining parts form a flat, leaf like photosynthetic organ known as 'Phylloclade'. In *Opuntia* species and *Muehlenbeckia*, the stems become flat and take up the photosynthetic function and are termed as 'Platyclades'. If the platyclades look like leaf as in *Ruscus*, they are called 'Phylloclades'

Morphologically and anatomically the leaf is the most variable plant organ. The foliage leaf usually shows its specialization as a photosynthetic organ by the expanded flat form of its blade, the lamina. The blade may be attached to the stem by petiole, or the leaf may be without a petiole. If the base of either a sessile or a petiolate leaf encircles the stem, it is referred to as a sheathing base. Sometimes the sheathing part of the leaf is developed into a conspicuous leaf sheath. Plant with multilacunar nodes have sheathing bases. Outgrowths at the base of the leaf, the stipules, are often present in the leaves associated with trilacunar nodes. The leaf surface which faces upwards and is closer to the internode above it is called as the adaxial, ventral or upper surface and the opposite surface as the abaxial, dorsal or lower surface.

HISTOLOGY OF ANGIOSPERM LEAF

Histologically the leaf is composed of three types of tissue systems: Dermal, ground tissue system and vascular.

Epidermis: Epidermis forms the dermal tissue system of leaf. The epidermis constitutes the outermost layer of cells of the leaves. Functionally and morphologically the epidermal cells are variable. Both upper and lower epidermis are usually uniseriate composed of compactly arranged cells. However in certain plants a multiseriate epidermis may develop as a result of periclinal divisions of the protodermal cells, e.g., Moraceae, certain species of Begoniaceae, Piperaceae, Chenopodiaceae. In *Ficus,* the upper epidermis is multiseriate whereas in *Nerium* both upper and lower ones are multiseriate. In *Begonia,* a succulent plant, the upper epidermis is three-layered while lower one is two-layered.

Epidermal cells: Various types of epidermal cells can be distinguished in different plants. These include the ordinary cells of the epidermis; single cells or groups of cells with special structure, form or content; cells connected with stomata; and epidermal appendages termed as trichomes.

The ordinary epidermal cells of the epidermis vary in shape, size and arrangement, but these are always closely attached to form a compact layer devoid of intercellular spaces. Many epidermal cells are tabular and in the leaf blade of dicotyledons the anticlinal walls are mostly sinuous. In the leaves of many monocotyledons, the epidermal cells are elongated. The external wall of the epidermal cells of certain leaves is raised in the form of papillae.

The epidermal cells wall differs in thickness. Some cells are thin walled, while in others the outer or the outer and inner periclinal walls are thicker than the anticlinal walls. Outside the outer cellulose wall there is usually a layer of pectin. Cutin is usually present in the outer walls of the epidermal cells and also constitute a special layer, the cuticle, on the outer surface of the cell wall. The cuticle possesses in some genera like *Eucalyptus*, a complex and species specific ornamentation which can be used for diagnostic purposes. The cuticle is of varying thickness in different plants growing in dry habitats. The surface of the cuticle may be smooth, rough, ridged and furrowed. Lignified epidermal walls are found in the leaves of the Cycadaceae, needles of conifers, in the strips of epidermis above the bundles of sclerenchyma, in the leaves of Poaceae, Juncaceae and Cyperaceae, certain species of *Eucalyptus*, *Quercus*, *Laurus* and *Nerium*.

Bulliform cells: Some epidermal cells found in groups in the leaves of certain monocots, especially grasses (maize, sorghum, sugarcane etc.) are comparatively larger, thin walled and highly vacuolated. These are called Bulliform (meaning bubble like) cells. These are commonly found in the upper surface of leaves, sometimes in lower surface also. Bulliform cells are mainly water containing cells, poor in solid and chlorophyll contents. Rarely tannins and crystals may be found. In cross sections, these appear as a fan-like band because the median cell being largest in size. There are three views regarding their function. (a) These are concerned with the unrolling of developing leaves by undergoing sudden and rapid expansion bringing about unrolling of lamina. (b) These are supposed to play an important role in hygroscopic opening and closing movements of mature leaves, due to changes in turgor. They are also called motor cells by some scientists. (c) These are concerned only with water storage and have no other functions.

Lithocysts are another specialised epidermal cells containing cystoliths and are found in the members of Acanthaceae, Moraceae, Urticaceae and Cucurbitaceae. Secretory cells which contain the enzyme myrosin are known as the myrosin cells and occur sometimes in the epidermis of members of Brassicaceae. Large idioblastic cells containing mucilage occur in the epidermis of Lythraceae. Very large sac-like cells which intrude into the mesophyll are found in *Cleome aspera*. Secretory cavities may sometimes develop from epidermal cells, e.g., in *Psoralea*. Long pipe-like cells filled with tannin substances appear in the epidermis of some species of *Saxifraga*. Silica cells may be present in epidermis as in certain grasses.

The continuity of the epidermis is interrupted by minute openings. These are intercellular spaces each of which is limited by two specialized cells termed as the guard cells. These cells, together with the opening between them, constitute the stoma. In many plants subsidiary or accessary cells can be distinguished. These cells differ morphologically from the typical epidermal cells and they constitute two or more cells bordering the guard cells to which they are apparently functionally connected. The stoma together with the subsidiary cells, if present, is termed the stomatal apparatus or the stomatal complex. The subsidiary cells usually develop from the protodermal cells adjacent to the stomatal mother cells, but they may also develop from the sister cells of the mother cell. Stomata are usually found on the aerial portions of the plant and especially on the leaves. If the stomata are present only on upper surface, the leaf is called epistomatic as in *Nymphaea*; if only on lower surface it is hypostomatic as in mango, oleander and banyan, and if on both the surfaces it is amphistomatic as in isobilateral leaves. Stomata are surrounded by kidney-shaped guard cells as in most dicots or dumb-bell shaped (double headed) guard cells as in grasses. In the broad leaves of dicotyledons the stomata are scattered. In the narrow elongated leaves of monocotyledons and conifers, the stomata occur in rows parallel to the long axis of the leaf. The stomata may be on the same level as the other epidermal cells as in mesophytic leaves or they may be located above the surface of the epidermis (raised stomata) or may be sunken as in *Cycas, Pinus* and *Allium*. Several stomata may appear in a depression called stomatal crypt e.g., *Nerium*. The size of the aperture between the guard cells increases or decreases as a result of turgor changes in the guard cells. As a result of unequal thickness and consisting of the walls the guard cells undergo changes in form with the increase in volume. These changes result in the opening of the stoma. In most cases the thinnest wall is that closest to the subsidiary cells and is termed the back or dorsal wall.

Protruberances of the guard cell wall may be present above or both above and below, the stomatal aperture. In cross section these protruberances appear as horn-shaped ledges. The outer ledge delimits the front cavity above the aperture, and the inner ledge delimits the back cavity which abuts on the sub-stomatal chamber.

The guard cells in Poaceae and Cyperaceae are different in form. They are elongated and dumb-bell shaped. The ends of these guard cells are expanded and thin walled, while the middle portions are elongated and thick walled and the cell lumen is narrow. As a result of turgor increase in this type of guard cell, the expanded ends swell and as a result push apart the middle elongated portions of the cells. The types of stomata have been given elsewhere in this book.

Below the stomata and directed inwards to the mesophyll are the large intercellular spaces which are termed as substomatal chambers.

The details of tricomes have been given elsewhere in this book.

Mesophyll

The mesophyll forms the bulk of internal tissue of leaf extending from upper epidermis to lower epidermis. In Dicotyledons two types of parenchyma can be distinguished in the mesophyll: Palisade parenchyma and spongy parenchyma. The palisade parenchyma consists of cells elongated perpendicular to the surface of the blade. The cells of the typical palisade parenchyma are elongated and in cross section of the leaf they are rod-shaped and appear to be arranged in rows, while in a section parallel to the leaf surface these cells are found to be rounded and seperated or only slightly attached to one another. The amount of palisade varies in different plants which also correlates with external environment. The xerophytic plants have well developed palisade tissue. In plants of the temperate regions characterized by abundant available water in the soil (mesophytic habitat), the palisade parenchyma is usually located on the upper (adaxial) side of the blade, the spongy parenchyma on the lower (abaxial). A leaf with such structure is called bifacial or dorsiventral. If the palisade tissue occurs on both sides of the leaf, as is common in the leaves of xerophytic habitats, the leaf is called unifacial, isobilateral or isolateral, e.g., species of *Atriplex, Centaurea, Artemisia, Myoporum*. In some plants, however, the palisade parenchyma is found only on the abaxial side of the leaf, e.g., *Thymelaea*. In *Rosa* and *Lawsonia*, the palisade cells are arranged in a single layer while in *Musa, Mangifera* and *Ficus* the palisade parenchyma is two to three-layered. In *Polyalthia* it is 2-layered. Many layered palisade parenchyma is seen in *Malus* and *Nerium*. In centric leaves such as *Allium* and *Juncus*, the palisade occurs all around the periphery of leaf. In a number of Dicotyledons such as *Caltha, Anemone* and *Aconitum*, and monocotyledons such as *Bambusa* and *Arundinaria*, the palisade cells have peculiar folds projecting into the cell cavity and these are called arm-palisade. In certain members of Piperaceae and Gesneriaceae, the palisade cells appear branched due to the presence of large lobes.

The spongy parenchyma consists of cells of various shapes, frequently irregular, with branches extending from one cell to the other. Limitation of such connections to the ends of the branches gives the spongy parenchyma an appearance of a three dimensional net, with the meshes enclosing the intercellular spaces. With regard to the connections between adjacent cells, the spongy parenchyma tissue has dominantly horizontal continuity parellel with the surface of the leaf, whereas the palisade parenchyma is continuous mainly in the direction perpendicular to the surface of the leaf.

In many monocotyledonous leaves, such as grasses, mesophyll tissues are not differentiated into palisade and spongy parenchyma and there occur only one type of cells. These cells contain abundant chloroplasts and intercellular spaces. In *Eucalyptus* and *Atriplex*, the mesophyll is entirely composed of palisade cells. Because of the shape and arrangement of the palisade cells, the chloroplasts are so placed to enable the maximum utilization of light. Another factor which increases the photosynthetic efficiency is the presence of well-developed intercellular spaces present in the mesophyll which facilitates rapid gaseous exchange. The large surface areas of the mesophyll cells are brought into contact with air. Together these surface areas are termed as 'the internal surface area' of the leaf and is distinct from the 'external surface area' of the leaf. The ratio of the volume of the intercellular spaces to the total volume of the leaf varies between 77:1000 and 713:1000. The specialization of the palisade tissue that results in more efficient photosynthesis is brought about not only by the increased number of chloroplasts in the cells but also by the dimensions of its free surface area. The intercellular spaces of the mesophyll usually develop schizogenously, but in certain plants, the development may also be lysigenous i.e., by the disintegration of group of cells, e.g., water and marshy plants and banana leaf.

Sometimes various types of idioblasts are found in mesophyll region. Presence of lithocysts in *Ficus*, oil glands in *Citrus* and laticifers in *Calotropis* are few such examples. In certain plants, stellate cells and trichosclereids are present to provide mechanical strength.

Secretory structures, which participate in the secretion of water or other substances, are a common feature of leaves. Many of them are of epidermal nature. The substances produced may be excreted from the cells or they may be retained only to be released upon the disintegration of the cells.

Glands are present on many foliage leaves. These structures consist of a mass of dense parenchyma cells in which terminates a vascular bundle. The parenchyma is covered by a glandular epidermis, the cells of which are mostly elongated in a direction at right angles to the surface of the gland, and have dense cytoplasm and large nuclei. The glands in different organs or parts of an organ of the same species may secrete different substances. In *Rumex persica* and other related species the glands on the laminar teeth secrete a bitter resinous substance, while those on the petiole secrete nectar. Nectariferous glands are present on the petioles of many plants like *Passiflora, Ricinus* and *Impatiens*.

Another type of the secretory structure is the essential oil cavity which occurs characteristically in the mesophyll of the leaves of *Citrus* and the secretory cavity of *Gossypium*. These are examples of lysigenous cavities in which the secretions formed in the cells are released after the disintegration of the cells, the lysis of which forms the cavity. Secretory cavities occurring in mesophyll of *Eucalyptus* develop lysigenously according to some authors and schizogenously according to others. Other types of secretory structures common in leaves are the schizogenous resin ducts of the Asteraceae, Anacardiaceae and Coniferales; laticifers of *Euphorbia*, mucilage cavities of Sterculiaceae, Malvaceae, Moraceae etc.

In leaves secretory substances may also be found in idoblasts. These secretory substances are classified according to the substances secreted. The secretory cells of the Lauraceae, Simaroubaceae and Onagraceae contain oils. Cells containing the enzyme myrosin are found in Brassicaceae, Capparaceae, Resedaceae, Tropaeolaceae and Moringaceae. Cells with a resinous content are found in the Meliaceae, Euphorbiaceae, Rutaceae and Rubiaceae. Tannin containing cells are common in Anacardiaceae, Annonaceae, Crassulaceae, Ericaceae, Euphorbiaceae, Buxaceae, Polygonaceae, Rosaceae, Tamaricaceae and Fabaceae. Cells with mucilaginous contents are found in Rhamnaceae, Malvaceae, Chenopodiaceae, Rubiaceae and many monocotyledonous plants. Idioblasts with different types of crystals and cystoliths (*Ficus*) are present in the leaves of different species.

Special secretory structures are the hydathodes which secrete water in the form of drops from within the leaf. This phenomenon is called guttation. Hydathodes secrete water which is brought to the surface by the terminal tracheids of the veins. The water passes through the intercellular spaces of the loosely packed parenchyma of the hydathode which is devoid of chloroplasts and is known as epithem. The intercellular spaces open to the exterior by special pores, which are of stomatal origin and which remain permanently open, e.g., *Brassica*, plants belonging to Gramineae, Crassulaceae and Saxifragaceae.

STRUCTURE OF THE PETIOLE

A similarity exists between the tissues of the petiole and those of the stem. The epidermis of the petiole is continuous with that of the stem. The parenchyma cells of the petiole, like those of the cortex, contain only few chloroplasts, compared to the cells of the lamina. The supporting tissues of the petiole are collenchyma and/or sclerenchyma. The vascular bundles of the petiole may be collateral (*Ligustrum*), bicollateral (*Nerium*) or concentric (certain Pteridophytes and many dicotyledons). In many species the phloem is accompanied by groups of fibres. The arrangement of vascular tissues in the petiole differs in different plants. In a cross section of the petiole it may appear as a continuous or an interrupted crescent (*Olea, Nicotiana, Nerium*) or as an entire or interrupted ring (species of *Ricinus, Quercus, Citrus*) or as a ring with additional internal and external bundles (*Vitis, Platanus, Robinia, Pelargonium*). An arrangement of many scattered bundles is seen in *Rumex* and many monocotyledons. If there is a single collateral bundle in the petiole, the phloem is found on the abaxial side and if the bundles are arranged in a ring the phloem is external to the xylem on the periphery of the ring.

In some plants, e.g., *Mimosa* and *Albizia* of the Leguminosae the petioles or petiolules have pad like swellings termed pulvini (pulvinus). These are capable of bringing about movement of leaves or leaflets. The pulvinus contains a large amount of parenchyma and its surface is usually wrinkled. The closure and opening movements of the leaflets may be stimulated by environmental factors and are brought about by turgor changes in the motor cells of the pulvini.

Vascular System

The midrib in most dicotyledons leaves have large bundle with adaxial xylem and abaxial phloem. This bundle is conjoint, collateral and open. A thin strip of cambium is present between xylem and phloem. In *Liriodendron* and *Vitis* the midrib region has several bundles arranged in a ring while in *Helianthus* several, but irregularly distributed bundles are found. The large and medium-sized veins contain vessels and sieve tubes. In the smallest veins the tracheary elements are tracheids with annular and spiral wall thickenings. The larger veins in some dicotyledons like *Ficus religiosa* show a limited amount of secondary growth as a result of activity of cambium. In monocotyledonous leaves the bundle is conjoint, collateral and closed.

In the small or minor veins the parenchyma cells in contact with the sieve elements and tracheary elements constitute transfer cells. Some of these cells are believed to be concerned with short distance translocation between mesophyll and sieve elements and others with an exchange of solutes between the phloem and xylem. The specialized transfer cells develop wall protuberances which increase the internal cell surface.

Bundle Sheath

The large veins are surrounded by much parenchyma which is poor in chloroplasts. The smaller veins, also, are usually surrounded by a layer of tightly packed parenchyma cells; such a layer is termed the bundle sheath. In dicotyledons, the cells of the bundle sheath are usually elongated in a direction parallel to the vein. Sometimes, however, for example in *Atriplex halimus*, the cells are more or less cubical. The cells of the bundle sheath are thin-walled. They may contain as many chloroplasts as the mesophyll cell, or they may be devoid of them. Sometimes crystals may occur in the bundle sheath cells. Bundle sheaths extend to the ends of bundles so that the vascular tissue is rarely directly exposed to the intercellular spaces. An exception is found in those hydathodes in which the xylem elements at bundle ends release water into the intercellular spaces. Bundle sheaths may be parenchymatic or sclerenchymatic and the two kinds of sheaths may be combined. Suberization of cell walls in bundle sheaths recorded in several species indicates that these cells might function as an endodermis.

In many dicotyledonous leaves such as the species of *Styrax* and *Quercus*, the bundle sheaths are connected with the epidermis by panels of cells resembling the bundlesheath cells. These panels are called bundle sheath extensions. Bundlesheath extensions have a conduction function in leaf, they conduct from the bundles to the epidermal cells. The latter are very closely connected laterally and therefore function to conduct in a plane parallel to the leaf surface. Water may also move from the bundle sheath and its extensions through the mesophyll to the cells lining substomatal chamber.

Bundle sheaths also occur in the monocotyledons, especially in the grasses in which two types are distinguished, one- or two-layered. The outer sheath when two layers are present, or the single sheath where there is only one, generally consists of parenchymatous cells with thin walls. The cells of this sheath contain chloroplasts which are usually smaller than those of mesophyll. The cells of the parenchymatous sheath may store starch and form a "starch sheath". The cells of the inner sheath, the mestome sheath, are smaller in cross section, their walls are thickened and contain suberized lamellae. The mestome sheath is thus analogous to an endodermis. Single sheaths are generally characteristic of panicoid grasses (Fig. 14.1 A) and double sheaths (Fig. 14.1 B) of festucoid grasses.

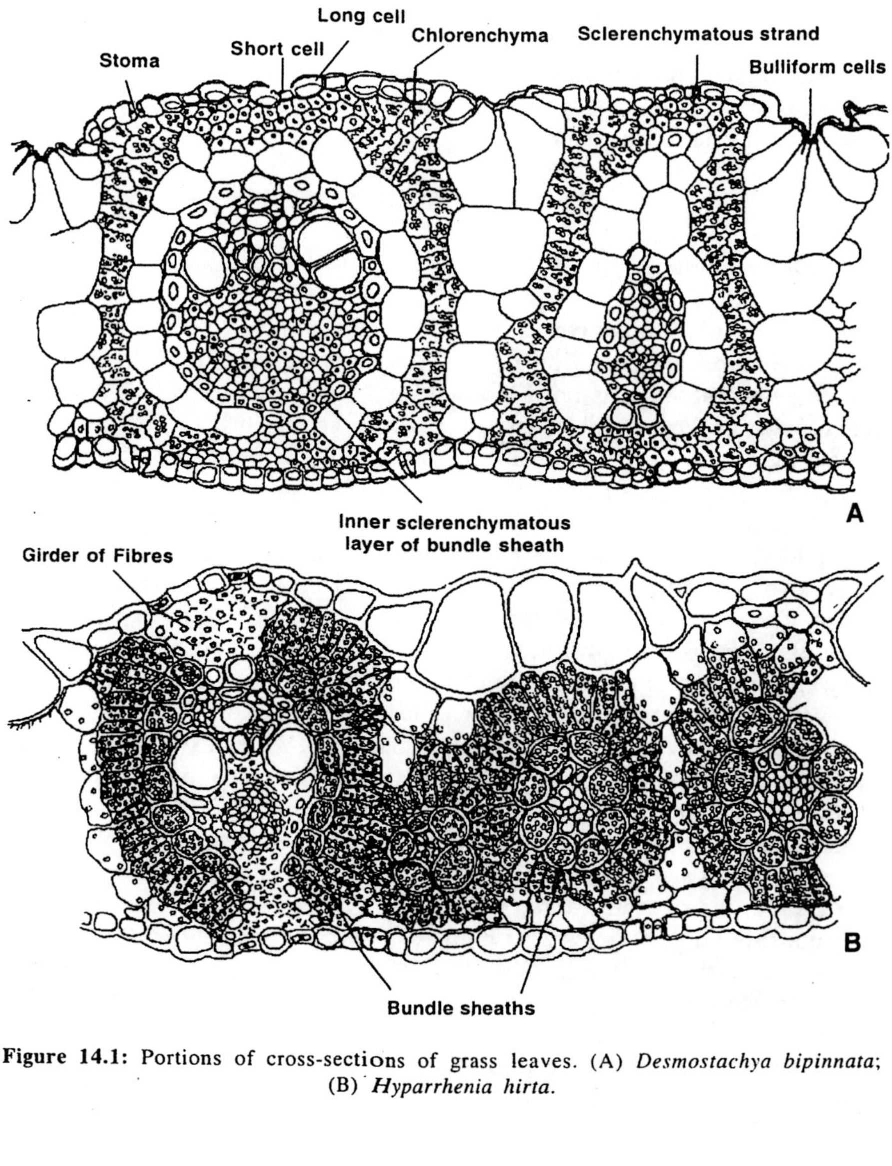

Figure 14.1: Portions of cross-sections of grass leaves. (A) *Desmostachya bipinnata*; (B) *Hyparrhenia hirta*.

Supporting Tissues of Leaf

The epidermis itself, due to its compact structure, strength of the cuticle, occasional thickness of its walls or deposition of the silica, gives support to the lamina. Collenchyma is usually present close to the larger veins immediately below the epidermis and also on the leaf margins in the dicotyledonous leaves. The bundle sheath extensions may also be collenchymatous. Additionally sclereids are also present in the mesophyll of many dicotyledons. The large and medium sized veins are accompanied by groups of fibres in many plants. In monocotyledonous leaves the vascular bundles are accompanied by many fibres. In Poaceae and many other monocotyledons the fibres form girders on one or both sides of the bundles.

LEAF DEVELOPMENT

Leaf Determination

The primordium which gives rise to leaf, can also give rise to branch of the stem, the inflorescence, a flower or in some cases, even thorns. So the fate of the organ primordium is not specified until it has emerged. It is shown by interposing an impermeable barrier between the leaf primordium and shoot apical meristem. This results in the growth of fewer leaves and more lateral branches. It appears that there is an important stage immediately after the primordium emerges in which its fate is not decided. In presence of signals from SAM, the primordium gives rise to leaf, when the signals from SAM are absent (due to impermeable barrier) the primordium may develop into a shoot. The timing of the above process can be investigated by explaining the individual primordia at different timings after its emergence and see whether they are able to give rise leaf tissue in cultures. The results of this experiment showed that plants vary in the timing of leaf determination. But in most plants it occurs soon after emergence of the primordium. By this, it is clear that leaf determination requires signals from SAM and this results in the polarization of dorsiventral axis. The dorsiventral polarity is seen in *phantastica* mutant. This mutant is identified in both *Antirrhinum* and *Arabidopsis*. The upper leaves in mutant plant are ventralized and form needle like structures, which resemble stem in their organization. The PHANTASTICA gene is very important component of dorsalizing signal in the beginning of leaf development.

Leaf Development

A leaf is initiated by cell divisions in the peripheral region of the shoot apex, more or less far below the distal zone of the apical meristem. However, the number of cell layers which divide and their position on the apex varies in different plants. In many grasses, the leaf initiation starts with periclinal divisions of the surface layer of the apex (outermost layer of tunica) and the cells of the layer immediately below it. So here, the main leaf primordium originates from the outermost layer of the shoot apex. In dicots and *Tulipa* (monocot), the first periclinal division occurs not in the surface layer of the apex, but one or more layers below it. Hence the surface layer gives rise only to the protoderm of the young leaf. But it is difficult to determine accurately the part played by each layer. The periclinal divisions are soon followed by anticlinal divisions in the protoderm and the layers beneath it. In some taxa, the protoderm cells also divide periclinally. The initial divisions give rise to lateral protrusion, often called 'leaf buttress' on the shoot apex. This leaf buttress consists of a protoderm layer, an inner mass of ground meristem and a procambial strand which develops acropetally from the nearby procambium of the stem.

The leaf primordium extends upward from the initial protrusion as a bladeless conical or peglike protuberance referred to as the leaf axis (Fig. 14.2 A). The primordium is dorsiventral as it emerges. In subsequent stages, the primordium increases in length and width and by growing more actively on the abaxial side than on the adaxial side, develops a curvature toward the shoot apex.

The increase in width of the leaf includes the spread of the leaf base laterally so that the apical meristem is encircled to a greater or lesser extent depending on the degree to which the mature leaf ensheaths the stem. The encircling part of the leaf base may give rise to stipules or to a sheath as in the monocotyledons.

The growth concerned with the lateral extension of the primordium becomes localized along two margins of the leaf axis and the primordium differentiates above the base into a midrib and two panels of the leaf blade (Fig. 14.2 B). Lateral veins become blocked out in the form of procambial strands as the blade continues to widen and, in petiolate leaves, a petiole develops between the blade and the leaf base (Fig. 14.2 C). The growing blade may spread side ways from the midrib or bend toward the adaxial side of the leaf (Fig. 14.2 D) so that it encloses the apical meristem and the younger leaf primordia. The midrib often increases in thickness on its adaxial side by orderly periclinal divisions within a region constituting the adaxial meristem. The petiole also increases in thickness by means of an adaxial meristem.

The early growth of a leaf is commonly divided into apical and marginal growth, the first concerned with the elongation of the primordium, the second with the lateral extension that produces the two panels of the blade. Thus, the growing primordium has an apical meristem at the tip and two marginal meristems along opposite sides of the axis (Fig. 14.2). Each meristem is at least two cells in depth.

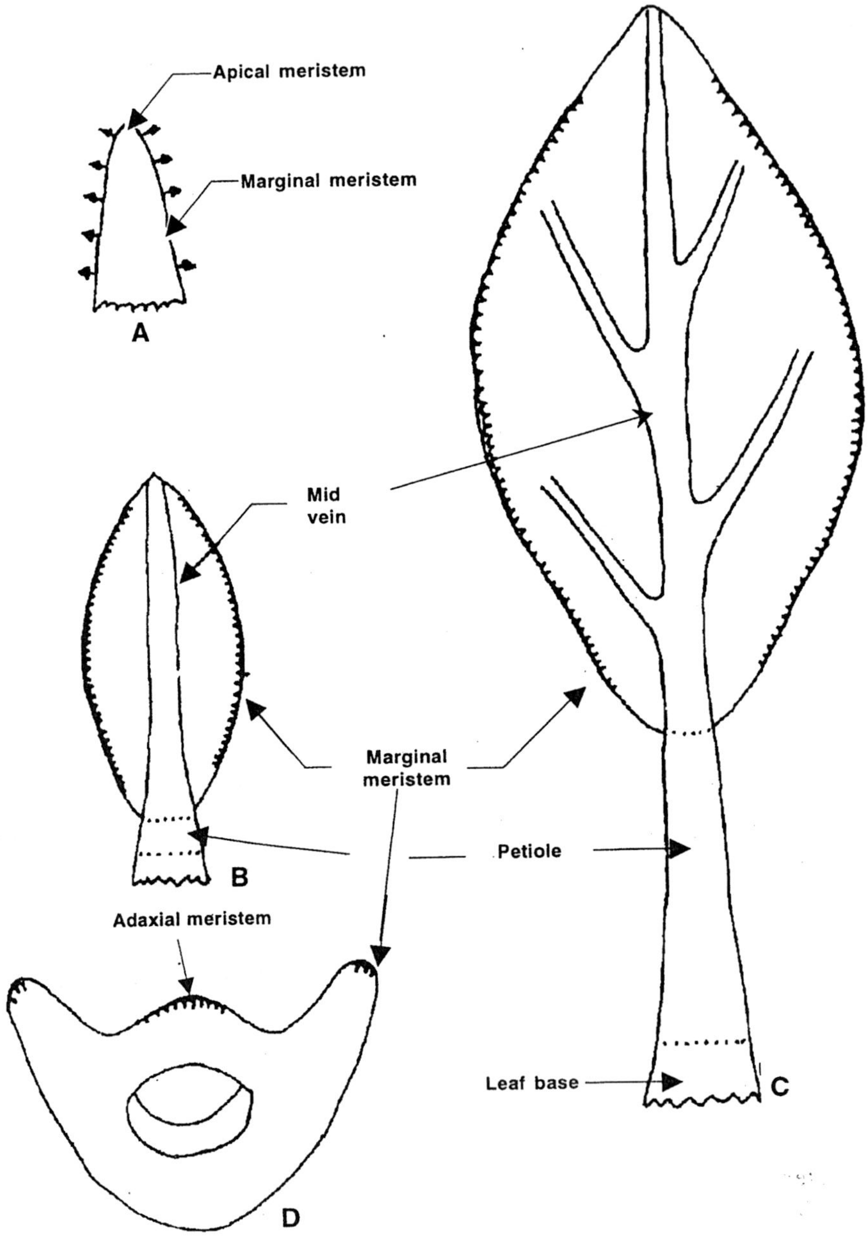

Figure 14.2: Diagrams interpreting growth of a dicotyledon leaf. (A) Undifferentiated leaf primordium (leaf axis) without blade and petiole; arrows indicate direction of growth; (B, C) Two stages of growth of blade; (D) Cross section of blade showing position of adaxial and marginal meristems.

The new cells that are added to the different layers of the lamina originate from rows of marginal initials and submarginal initials. The marginal initials are the cells of the outer most layer on the margins of the young lamina. Generally in the angiosperms these initials divide only anticlinally and so add new cells to the abaxial and adaxial protoderm. The submarginal initials undergo divisions in various planes. These cells give rise to new cells which are added to the inner layers of the young lamina. In Fig. 14.3 and 14.4, it can be seen how, in different plants, these tissues develop from the cell layers of the young lamina.

In pinnately and palmately compound leaves lateral leaflets develop from the adaxial marginal meristem of the axis of the young leaf as two rows of papillae. In certain plants the order of development of the leaflets is acropetal, e.g., *Carya*, but in many other plants the order is basipetal. The development of lobed leaf is a result of differential activity of the leaf margin which starts early in ontogenesis. In the regions of the future lobes meristematic activity is considerable, while in the regions of future sinuses there is little or no activity.

The derivatives of cells at the apex and along the margins of the leaf divide in their turn. These intercalary divisions, which are distributed at random, make the major contribution to the increase in size of the leaf. The intercalary meristematic activity ceases earlier at the apex than at the base so that the leaf tissues differentiate and mature in the basipetal direction.

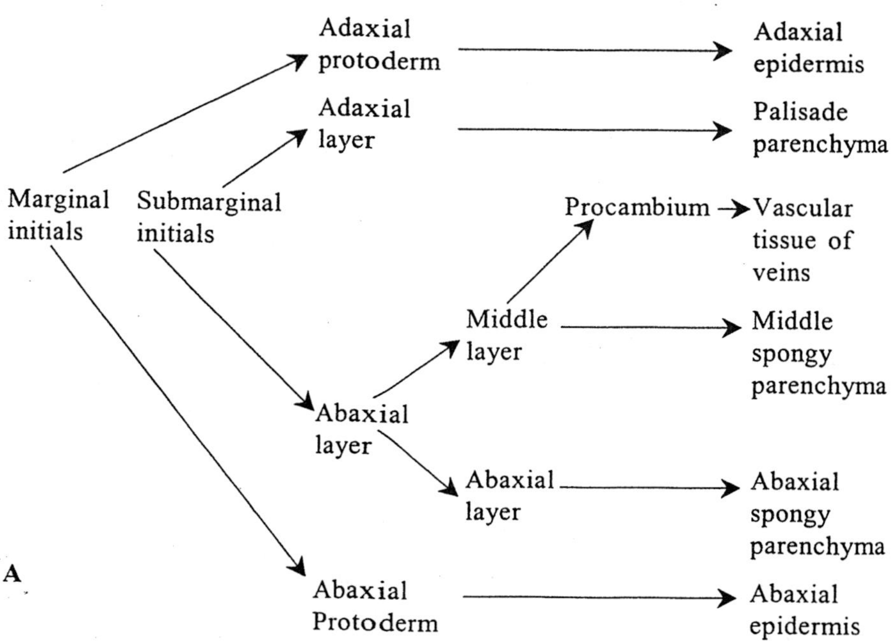

A

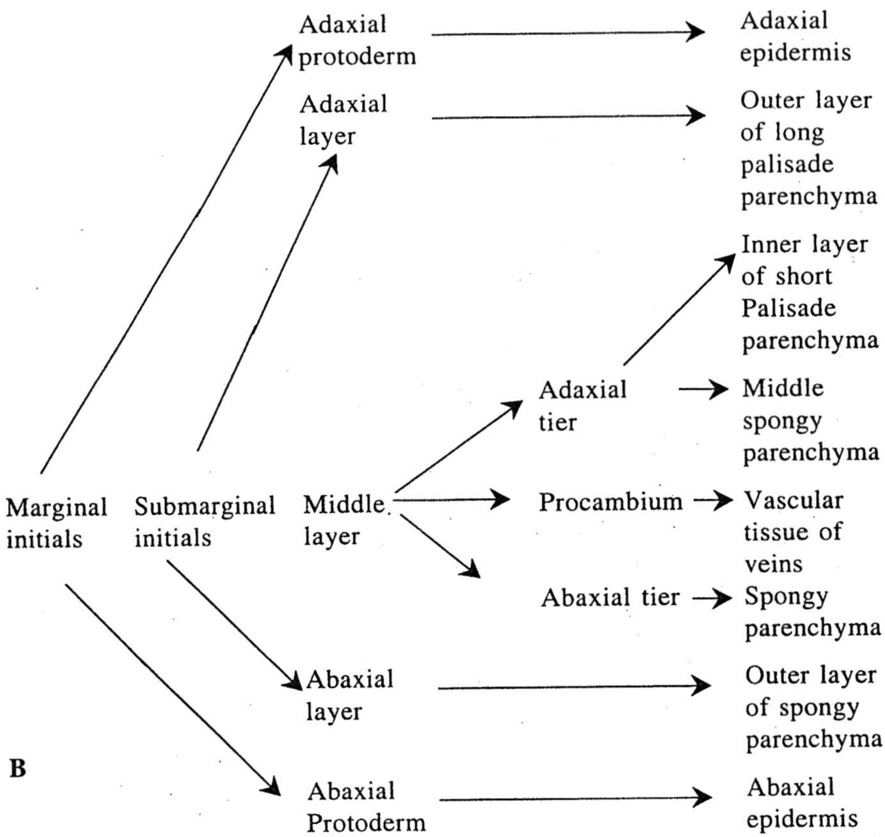

Figure 14.3: Diagrammatic representation of leaf histogenesis (A) *Carya buckleyi*; (B) *Pelargonium zonale*.

Moreover, cell divisions may be less numerous in the apical region than farther below, a difference that results in a greater width of the leaf toward the base. The basipetal order of maturation is particularly clear in the long narrow monocotyledon leaves in which growth becomes conspicuously localized at the bases of leaf blade and sheath zone is referred to as intercalary meristems.

DEVELOPMENT OF STOMATA

Stomata are essential to the life of terrestrial plants. There is usually a stomata free zone surrounding each stoma in normal plants. This intervention of subsidiary (when they exist) or epidermal cells between stomata is thought to provide a pool of ions for the guard cells and optimize the access of the underlying mesophyll cells to the external environment. Two features that

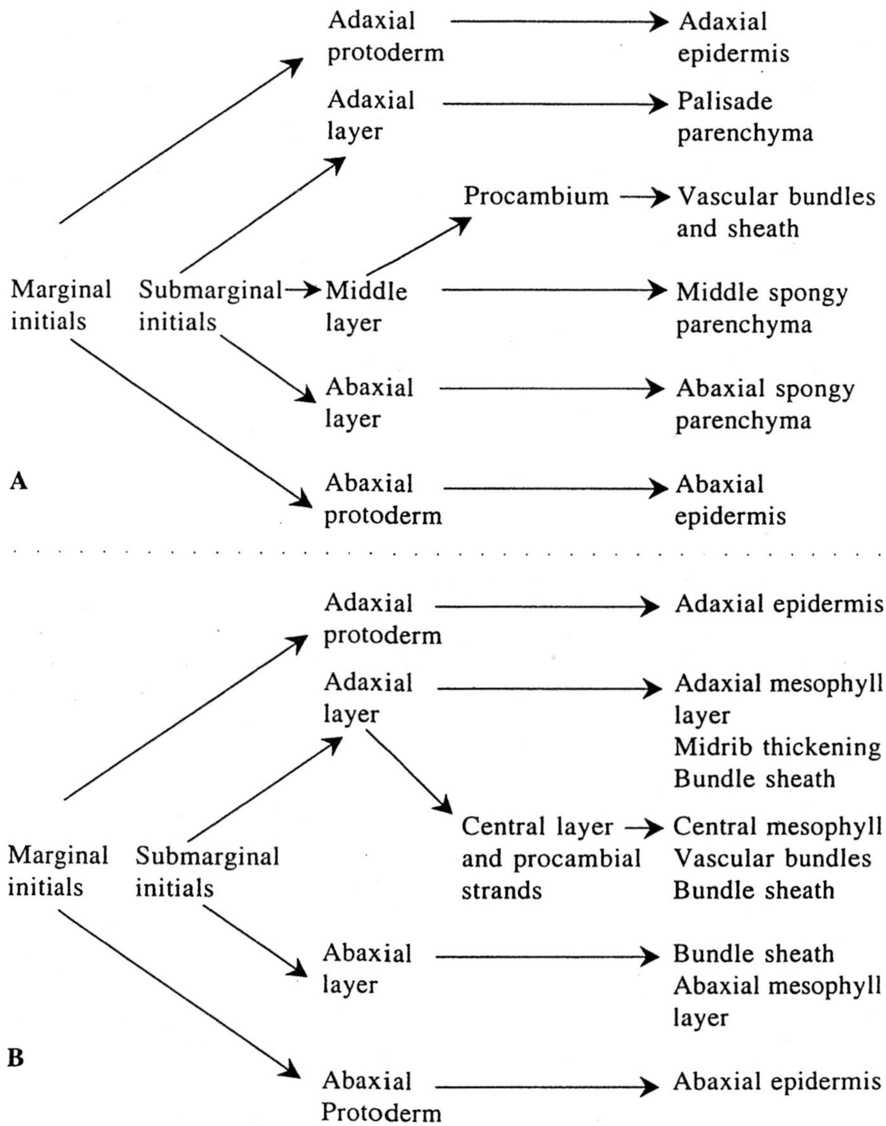

Figure 14.4: Diagrammatic representation of leaf histogenesis. (A) *Nicotiana tabacum*; (B) *Oryza*.

influence stomatal patterning are: internal anatomy (cell type, position of the vasculature), and contact between stomata. Little is known about that why stomata rarely develop over the vasculature. Certain underlying cell types seem to inhibit development of stomata over vasculature. Stomata also rarely develop side by side as there is a control over stomatal patterning.

Stomatal initials form by asymmetric cell division of precursor cells such that the smaller cell becomes a stomatal initial. The stomatal initial can produce both the guard cells and ordinary epidermal cells. But the mother cell of the guard cells always undergoes symmetrical division to produce two guard cells. Cell-to-cell communication takes place leading to the co-ordination of the plane of cell division as well as the regulation of precursor cell activity. Additionally, the timing of when a cell is competent to become a stomatal precursor and where such precursors can initiate are highly controlled.

In both dicots and monocots, the cell divisions leading to the production of stomata are highly asymmetric. In monocots, the first asymmetric division results in the smaller, usually rectangular cell becoming guard mother cell, which divides symmetrically producing two guard cells. In dicots, the smaller cell is usually triangular and continues to divide after the surrounding cells have ceased to do so. Eventually it converts into guard mother cell and divides symmetrically to produce two guard cells. Since dicot stomatal initials continue to divide after the surrounding epidermal cells have ceased they have been termed meristemoid cells to emphasize their ability to continue cell division. In dicots there are two types, primary meristemoids and secondary (satellite) meristemoids, the latter produced by an asymmetric division of a neighbouring cell to the stomata. Primary meristemoid positioning appears to be random, with the exception that it obeys the rule that two stomata should not be adjacent. Secondary meristemoids and stomatal initials in monocots are placed in a highly regular fashion crucial to stomatal patterning. Primary meristemoids arise from protodermal cells. These cells undergo a series of asymmetric divisions that produce a small, triangular initial situated approximately in the middle of the future stomatal complex. Secondary (satellite) meristemoids arise once a primary meristemoid has completed development into a stoma.

Generally, the mother cell of the stomata divides symmetrically to fom two cells that differentiate into the guard cells. At first these cells are small and have no special shape but, as they develop, they enlarge and become characteristically shaped. During their development the middle lamella between the two guard cells swells and becomes lens-shaped and finally it disintegrates to form the stomatal aperture. Stevens and Martin (1978) suggested that, pore formation is initiated enzymatically. The separation of the two guard cells is brought about by osmotic force derived from starch hydrolysis. The mature guard cells are sunken or raised relative to the level of ordinary cells of the epidermis. The sinking or raising are brought about during the maturation of the guard cells. The development of stomata in the leaf continues for a relatively long period during the growth of the leaf.

Timing of stomatal development: In *Arbidopsis* at least, stomata develop after trichomes have formed.

Cell-to-cell communication during stomatal development: There is a wide range of signals traversing between cells thus coordinating stomatal development. Evidences for cell-to-cell communication are: (a) Inhibition of stomata formation above veins in the leaf, (b) Development of meristemoids besides another meristemoid or stomata is arrested, (c) The orientation of subsequent cell division in such a way to separate meristemoids when two happened to develop beside each other, and (d) the initiation of subsidiary cell formation in monocots.

Florin *et al.* (1952) recognized two types of stomata based on ontogeny: (a) Haplocheilic type: Guard cells arise by a single division of the stomatal initial, and the neighbouring cells become modified into subsidiary cells, (b) Syndetocheilic: Guard and subsidiary cells are derived from the same mother cell. Pant (1965) recognized three categories of stomata on the basis of development.

Mesogenous type: Guard cells and neighbouring cells, which may or may not differentiate as subsidiary cells have common origin, e.g., Rubiaceae, Brassicaceae. The initial of the stomatal complex is formed by an asymmetric division of a protodermal cell. Two divisions result in the partitioning of the initial cell into a mother cell of the guard cells and two subsidiary cells. One more division in the mother cell of the guard cells leads to the formation of guard cells (Fig. 14.5).

Perigenous type: Neighbouring or subsidiary cells do not have common origin with the guard cells, e.g., Cucurbitaceae and Nymphaeaceae, Poaceae (Fig. 14.7). In grass leaves and some other monocotyledons, the first step in the differentiation process is the symmetric division of certain protodermal cells. Before the cell divides, the nucleus becomes displaced in one direction and the vacuoles occupy the other end of the cell. The nucleus then divides. One of the daughter nuclei which is closer to the vacuoles becomes larger and more weakly staining than the other at the denser end of the dividing cell. The smaller distal nucleus divides later to form the two guard cells, but before doing so it induces neighbouring epidermal cells to divide. In these cells also there is denser cytoplasm adjacent to the guard mother cell, and the induced divisions are also asymmetrical. The daughter nucleus in the centre of the cell becomes larger and weaker staining than that near the guard mother cell, which becomes the nucleus of a subsidiary cell. After the formation of the subsidiary cells the guard mother cell divides to form the guard cells (Fig. 14.7 C–G).

Mesoperigenous type: At least one neighbouring or subsidiary cell and stomata are derived from the same mother cell (Fig. 14.6), e.g., Ranunculaceae and Caryophyllaceae. As a rule the precursor cell undergoes an asymmetric division to produce a smaller stomatal initial and larger ordinary epidermal cell. The smaller cell undergoes a division giving rise to a subsidiary cell

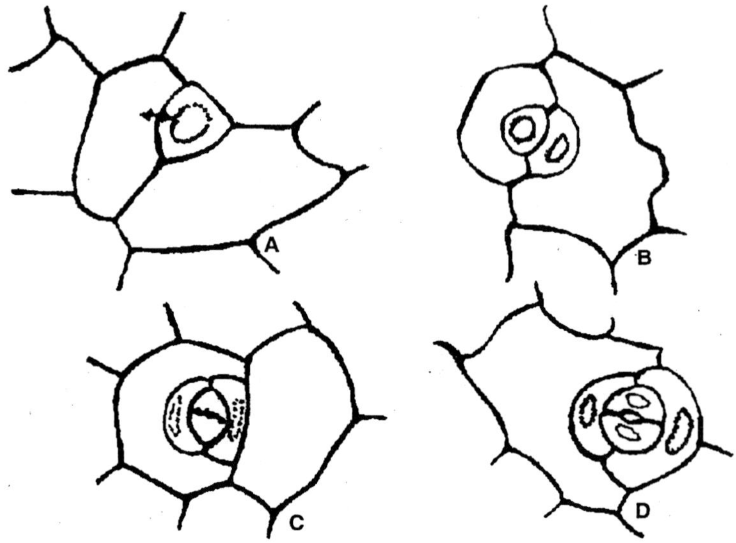

Figure 14.5: Development of stoma with mesogenous subsidiary cells in a leaf of *Thunbergia erecta*.

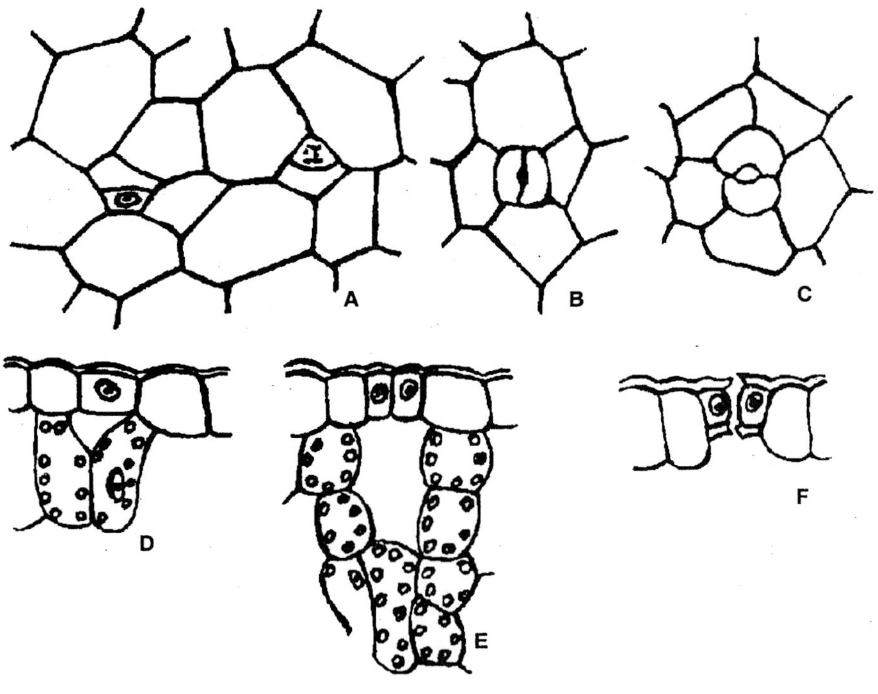

Figure 14.6: Development of mesoperigenous stomata in *Beta vulgaris*. (A–C) Surface view; (D–F) Sectional views.

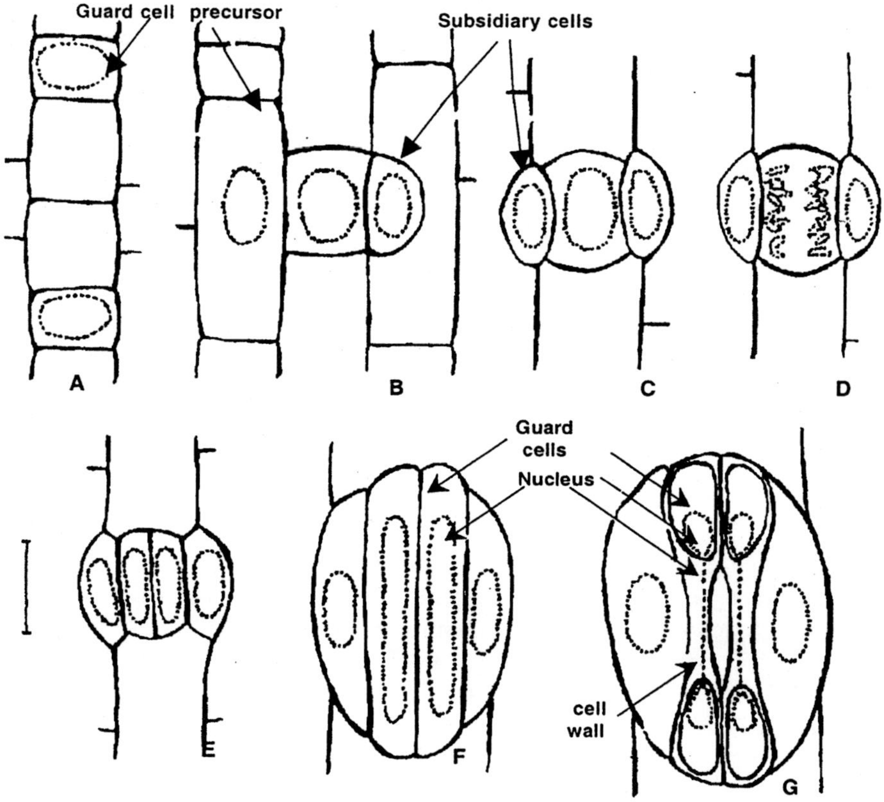

Figure 14.7: Development of perigenous stomata in *Avena sativa.*

and an initial cell. Then the initial cell undergoes a division at right angles to the subsidiary cell wall that is already formed and produce another subsidiary cell. Finally the initial divides to produce third subsidiary cell and mother cell of the guard cells. Then the mother cell of the guard cells divides symmetrically to produce two guard cells. As they develop, they enlarge and become characteristically shaped (Fig. 14.6).

GENETIC ANALYSIS OF STOMATAL DEVELOPMENT

There are two mutants identified that violate the rule that two stomates should not be in contact. The four lips (flp) and too many mouths (tmm) mutations show two different developmental mechanisms, tmm mutants have greater numbers of stomata in rosette leaves, cotyledons and abaxial sepal surface. Other tissues of tmm plants (inflorescence, stem, adaxial sepal, epidermis etc.) entirely lack stomata where as wild type plants possess them. The flower pedicel shows an abnormal stomatal frequency from none proximal to the stem to abnormally greater numbers near the flower base. Hence, TMM is thought to control the entry of protodermal cells into the stomatal developmental process and appears to have opposite effects on meristemoids depending on the tissue (compare rosette leaves with sepal epidermis). The tmm mutation appears to increase the formation of satellite meristemoids as well as to alter the polarity of cell divisions such that meristemoids can now form next to guard cells. The flp mutation can produce unpaired guard cells as well as result in stomatal twinning. Analysis of stomatal development in flp mutants suggests that stomata develop normally until the guard mother cell stage. At this stage the guard mother cell undergoes a symmetric division and the resulting daughter cells divide symmetrically again to produce two contiguous stomates. Hence flp may control guard mother cell identity, or regulate the number of symmetric divisions a guard mother cell can undergo.

DEVELOPMENT OF TRICHOME

A trichome has been defined as any protrusion from the epidermis. Accordingly conical-papillate petal cells, both glandular and non-glandular hairs, and surface glands are all classified as trichomes. Although some plants have single type of trichome, but often different types of trichomes are reported on the same plant. For example, tomato has six different types of trichomes. Trichomes play a role in protecting the plant from predators and the disease-causing organisms they carry. Trichomes mature basipetally.

Trichomes are differentiated from single cells originating in the epidermis. Although trichomes occur on different parts of the plant, the leaf trichomes are special because they branch characteristically and are evenly distributed across the leaf's surface. The majority of flowering plants produce multicellular trichomes. Although these trichomes must undergo processes of commitment, expansion and morphogenesis, which may, in principle, be similar to those of unicellular trichomes, they must also do this in the context of continuing cell division. Multicellularity allows trichomes to function as miniature organs, with differentiation within the trichome with regard to cell function. Many multicellular trichomes are glandular, developing a terminal gland, which may secrete a variety of compounds, such as alkaloids to poisonous substances. Tobacco leaves develop two types of multicellular trichome, described as short and long-stalked. The short-stalked trichomes have an eight-celled glandular head, and secrete nicotine, while the long stalked trichomes have a gland that produce diterpenoids.

Ontogeny: Trichomes develop from single epidermal cells in young leaf primordia and have a remarkable cell morphogenesis. The trichome cell differentiate while epidermal pavement cells are still dividing and it synthesizes more DNA at the same rate as they do, but cell division is arrested.

After three rounds of endoreduplication of nucleus, the cell expands outwards from the leaf surface as a single cone. The enlarging cell grows outwardly from the leaf surface, so its nucleus is pushed into a subapical location. When the volume of the trichome cell has increased tenfold, it branches. The nucleus enlarges again by endoreduplication, while the two newly formed branches grow at their tips. One of the branches elongates to form the main branch, then branches again. The nucleus then relocates within the cell so that it lies between the two branching points. The mature trichome rests on a small podium formed by a circle of 8–10 epidermal cells, around a layer of subepidermal cells. The epidermal podium cells differ in shape and size from other cells in the epidermis.

Trichome development is linked to leaf development. Trichomes emerge from the leaf primordium a distance of approximately four cells apart. As they grow and branch, the epidermal cells between them divide. In the end, approximately 30 epidermal cells separate neighbouring trichome cells. The branching of the trichomes is oriented according to the longitudinal axis of the leaf primordium. The first-formed branch points towards the base of the leaf. The latter branching main branch grows towards the leaf tip. The trichomes of a leaf primordium are developed to different degrees: at the leaf tip are trichomes that are already mature, while at the base some trichomes have yet to branch, and in the leaf centre there are trichomes that have one branch.

The pattern of trichome development: The pattern of trichome development in the *Arabidopsis* leaf epidermis is not random. There are two models to explain how trichome spacing is determined. The first suggests that precursor cells produce inhibitory signals that prevent neighbouring cells to develop into trichomes. The second suggests that, once a precursor cell is induced to become trichome it undergoes several divisions that separate it from other trichomes by a group of cells, all derived from the same precursor daughter cell. A model based on genetic analysis has been proposed to explain how events proceed to commit individual cell further to trichome differentiation while inhibiting neighbouring cells.

There are atleast 10 known genes that effect trichome development, of which we shall discuss 7. There are core protein types in trichoblasts (cells destined to become trichomes) and three in atrichoblasts. Two of these core protein types are the same in the two types of cells. The trichoblast stimulating complex is formed from GL1:GL3:TTG1(GL1 = GLABRA1; GL3 = GLABRA3 and TTG 1 = TRANSPARENT TESTA GLABRA1). The atrichoblast complex is formed from TRY:GL3:TTG1 (TRY = TRIPTYCHON). It is the competetion between TRY and GL1 to interact with the other proteins in the complex that decides trichoblast fate. The GLABRA1 (GL1) gene (glabrous = free from hair) which, when defective (gl1), results in the loss of trichomes on most surfaces and the TRY gene, are both MYB transcription factors. The GLABRA 1(GL3) protein and the ENHANCER OF GL3 (EGL3) protein both encode basic Helix-Loop-Helix (bHLH) proteins that interact as homo (= a temporary structure) to which TTG1 and GL1'TRY proteins bind. Neither the gl3 nor egl3 mutants have striking phenotypes but the double mutant is totally without trichomes. The TRANSPARENT TESTA-GLABRA1 protein is a member of both the trichoblast stimulating and trichoblast inhibiting complex. Mutants in ttg1, is a mutant that results, among other things, in the loss of trichomes from most surfaces of the plant, akin to gl1 mutations. An additional player that influence this competetion is CAPRICE (CPC), also a MYB transcription factor. CPC negtively regulates trichome initiating by binding to GL1, making it unavailable to participate in the trichome initiating complex. Both the trichoblast stimulating and trichoblast inhibiting complex bind to the promoter region of the GLABRA2 (GL2) gene. In trichomes, gL2 is up-regulated by the trichoblast stimulating cimplex (GL1:GL3:TTG1) and down-regulated by the trichoblast inhibiting complex (TRY:GL3:TTG1).

DIFFERENTIATION OF MESOPHYLL

The mesophyll differentiatiates from the derivatives of the marginal meristem after these derivatives have undergone the intercalary growth. The establishment of the characteristic differences between the palisade and the spongy parenchyma results from an unequal growth in the various layers of the leaf: there are differences in duration of cell division and cell enlargement and in the direction of cell expansion in the epidermis and in the various layers of the mesophyll. Cell division commonly ceases first in the upper epidermis and continues longest in the future palisade tissue.

The difference in amount of cell division between the upper epidermis and the palisade parenchyma may be illustrated by the numerical relationships between cells of the two tissues in young and old apple leaves. The ratios between the numbers and diameters of the epidermal and the palisade cells were 1:1 in young leaf. In the mature leaf the diameters of the epidermal cells are 3 to 4 times larger than those of the palisade cells, and there were 8 to 10 palisade cells to one epidermal cell. In *Xanthium* leaf, the upper epidermis and the palisade parenchyma differ in duration of cellular enlargement and rates of expansion. In the epidermis this rate is high in the horizontal plane but low in the vertical. The opposite relation characterizes the palisade tissue. While the division and the enlargement of the palisade cells remain correlated with the growth of the epidermal cells, the palisade cells are closely packed. When the growth of the two kinds of cells becomes differential, intercellular spaces develop in the palisade tissue. The palisade cells divide mainly by anticlinal walls. The palisade cells also elongate at right angles to the surface, and during the formation of intercellular spaces they separate from each other along the anticlinal walls. All these phnomenona are responsible for the characteristic appearance of the palisade, that is, of a tissue composed of orderly rows of elongated cells largely separated from each other along their anticlinal walls. The developmental relation between the lower epidermis and the spongy mesophyll in bifacial leaves appears to be somewhat variable. This epidermis may cease cell division before the spongy mesophyll but continue longer with cell enlargement or it may show cell division after this activity has stopped in spongy tissue. In both situations the epidermis grows in surface without formation of intercellular spaces, while the spongy parenchyma grows in a tangential plane by some enlargement of cells and by a loosening of contacts. Among the various tissue elements of the leaf, the epidermal hairs, the stomata, and the large veins complete their differentiation before the mesophyll. The stomata develop concomitantly with or after the development of the intercellular spaces in mesophyll.

DEVELOPMENT OF VASCULAR TISSUES

Vascular development in a leaf of a dicotyledon begins with the differentiation of procambium in the future midvein. The procambium differentiates acropetally in continuity with the leaf trace procambium in the axis. With the commencement of the development of the lamina the procambial strands of the large lateral veins and, later of the smaller veins begin to form gradually. The procambial strands of the small veins, which form in a basipetal direction develop mainly during the intercalary growth of the lamina.

In the Monocotyledons the procambium of the larger veins differentiates acropetally, that of the smaller veins alternating with the larger and of the transverse inter connections progresses basipetally.

STRUCTURE OF MODIFIED LEAVES

Xeromorphic foliage Leaves: The term xeromorphic is used to indicate plants that have anatomy and growth form which are adapted to desert conditions, particularly where the water is scarce. In addition, the desert soils are typically alkaline and are rich in minerals and sometimes toxic too. They are poor in organic materials which helps to hold water in the soil. Temperatures vary between extremely hot and severely cold on both daily and on annual basis. The plants meet these challenges in various ways. They germinate, grew, set seed and die within few weeks and complete their life cycle. So their adaptation to xeric conditions is their ability to complete their life cycle in a short span, when the desert is temporarily mesic after rains.

Plants that occur in xeric microhabitats are perennial and show well developed xeromorphic characters, such plants are called xerophytes. These xeromorphic characters are: thick walled epidermis and hypodermis covered by a dense, waxy cuticle, trichomes are abundant. These characters are advantageous in many ways, i.e., all these characters are effective in blocking excess sunlight. They are also good deterrents against insect feeding and egg laying. Salt glands are also present in many cases and they also help in protecting against feeding and sunlight as well as keeping salts accumulating in the protoplasm. Some of these characters are directly induced by the environment. Plants of *Prosopis* growing outdoors have cuticle ten times thicker than the plants that are grown in doors.

The surface area of the xeric leaves is greatly reduced for decreasing the transpirational surface area. Leaves are smaller, more cylindrical and are more succulent. To further reduce water loss, the stomata are sunken and are restricted to well protected crypts, e.g., *Nerium, Ficus.*

Inside the leaf too, the surface area is reduced, as the cells are packed together and are smaller. The spongy mesophyll may altogether lost, with only palisade parenchyma remaining. The vascular bundle sheaths are sclerenchymatous. Water storage cells are present as large mucilage cells, e.g, Cacti.

Submerged Foliage Leaves: A small number of angiosperms and a few gymnosperms have migrated back into aquatic habitats. These plants are quite variable ranging from the plants growing in marshy swampy areas to growing as floating unanchored plants and plants rooted to the bottom of the stream, pond or ocean, its leaves are fully submerged and plants are called hydrophytes, e.g., *Ceratophyllum, Vallisnaria, Sagittaria, Nymphaea.* As the water provides a uniform habitat, the anatomical features of water plants are less varied that the xerophytes. The factors influencing the water plants are temperature, air and the concentration and composition of salts in the water. The most striking features of the water plants are, the reduction of supporting and protective tissues, decreasing of vascular tissues especially xylem and the presence of air chambers.

The epidermis of water plants does not have a protective function, it plays an important part in the uptake of nutrients from the water and in gas exchange. The cuticle and the cell walls are very thin and the epidermal cells contain the chloroplasts. Immersed leaves are generally devoid of stomata or may possess vestigial stomata. They are present on the upper surface of the floating leaves. In the epidermis of the immersed surface of leaves of some water plants, small groups of cells appear which stain deeply with certin dyes. These cell groups are termed as 'hydrophytes' are thought to be structures which facilitate water and salt transport into and out of the plants.

In many water plants like *Ceratophyllum, Myriophyllum* and *Utricularia* the leaves are divided into narrow cylindrical lobes which greatly increase the area of contact with the water. In *Ranunculus aquatilis*, the floating leaves are entire while the immersed leaves are divided into narrow lobes.

In the leaves and stems of hydrophytes air chambers are present which are filled with gases. These chambers are intercellular spaces regular in shape and pass through the entire leaf. The air chambers are separated from one another by thin partitions and allow only the passage of gases but not water. Immersed hydrophytes contain very little or no sclerenchyma. However, strips of sclerenchyma are present along the leaf margins.

Halophytic Leaves: Halophyts grow in saline soil of water where there is abundance of salt in the soil. Typical examples of halophytes are *Chenopodium*, *Acanthus*.

Halophytes growing in marshy places near the sea-shore collectively form special vegetation called mangrove. Anatomically mangroves show several xerophytic characters. The epidermis is thick-walled and strongly cutinized. Sunken stomata are found in the epidermis. Mesophyll cells are usually compactly arranged.

Kranz Anatomy: Haberlandt (1918) distinguished in the leaf cross-section of certain plants a layer of radially oriented mesophyll cells which surrounds the vascular bundles (Fig. 14.1 B). He called this layer "Kranz" (wreath) and the type of leaf anatomy Kranz type.

The C_3 and C_4 plants show anatomical and ultrastructural differences with a high degree of consistency. In C_3 plants this sheath has few organelles and rather small choloroplasts so that at low magnifications cells appear empty and clear in striking contrast to the chloroplast-rich mesophyll. In C_4 plants the sheath has a high content of organelles especially mitochondria and its chloroplasts are deep green and large, commonly larger than the mesophyll chloroplasts. The sheath cell chloroplasts may be localized next to the outer tangential wall or on the opposite side of the cell. The walls of the sheath are somewhat thickened. When the vascular bundles of C_4 plants are enzymically isolated from the mesophyll the sheath is found to be tightly adhering to the vascular tissue, with its cells arranged in double helix. An extreme case of this ultrastructural dimorphism is seen in sugarcane where the bundle sheath chloroplasts lack grana. In other C_4 plants, grana are found in all the chloroplasts. Numerous large mitochondria are present in bundle sheath cells. Microbodies are present in bundle sheath cells than in the mesophyll cells. Another characteristic feature of C_4 chloroplasts is the occurrence or the peripheral reticulum (PR). The PR is generally more developed in mesophyll chloroplasts than the bundle sheath chloroplasts. The PR is involved in transport of materials between the chloroplasts and cytoplasm. It appears in the peripheral stroma as a series of anastomosing tubules contiguous with the chloroplast envelope.

C_4 photosynthesis has been recorded in Aizoaceae, Molluginaceae, Amaranthaceae, Portulacaceae, Nyctaginaceae, Zygophyllaceae, Asteraceae, Cyperaceae and Poaceae.

CATAPHYLLS

The cataphylls or bud scales function as a means of protecting the apical or axillary shoot tips during the times of dormancy. Annual plants do not have terminal resting buds and hence lack bud scales. Some of the tropical plants like *Hevea braziliensis, Mangifera indica* have resting buds with bud scales or cataphylls. Many temperate, perennial species have a means of protecting their apical regions from damage. There are several dangers to the meristems and leaf primordia within a bud. The most important among them is the dessication and insect predation. Winter winds absorb a great deal of moisture and the immature vascular tissues are not able to conduct water rapidly during winter. Further, the young stems and leaves are the richest sources of proteins, a favoured source of food for insects. To combat these dangers, the bud scales are small, sessile or have a very short petiole. The longitudinal growth of the bud scales are small, sessile or have a very short petiole. The longitudinal growth of the bud scales stops earlier than the foliage leaf of the same plant and the marginal growth starts sooner. The marginal growth of the bud is also less than that of the foliage leaf. But in conifers with needle leaves, the cataphylls are broader than the needles. The epidermis is thick walled or it forms a cork cambium and produces a thin, protective, corky bark. The cork is a good water retention device and is also indigestible by the insects. Resin, oils and waxes also are secreted. Secretion and cork formation occur only in the exposed portions of bud scales. Stomata are rare. The vascular supply is simple, because very little conduction is needed. The mesophyll is only spongy but palisade is also present in green bud scales. Secretory cells and cavities occur in those that produce pungent oils.

HYPSOPHYLLS

Hypsophylls or floral bracts which protect the developing inflorescences and their function is similar to cataphylls and are active during the formation of inflorescences. The environmental conditions that they shield against are milder than those faced by cataphylls. Hence the hypsophylls are not as resistant as cataphylls. Mostly they are green and photosynthetic. Anti-insect mechanisms are present in the hypsophylls, when insects are more abundant.

CHAPTER XV

THE FLOWER

The flower consists of an axis on which the rest of the floral organs are borne. That part of the axis that represents the internode terminated by the flower is termed the pedicel. The distal end of the pedicel is swollen to various extents and this portion is termed the floral receptacle or thalamus. The floral organs are attached to the receptacle. A typical flower has four types of organs. The outer most organs of the flower are the sepals which together constitute the calyx which is usually green and is found lowest on the receptacle. On the inside of the sepals is the corolla , consisting of the petals which are generally coloured. These two types of organs together form the perianth; however, sometimes one of them may be lacking. Within the perianth two kinds of reproductive organs are found: externally the stamens which together form the androecium and internally the carpels which form the gynoecium.

The floral initiation programme is activated in many plants by signals from the environment, especially day length, which connvert the leaf producing SAM into a flower producing inflorescence meristem. Floral organ formation and patterning occur in the determinate floral meristem, which arise from the indeterminate inflorescence meristem during floral initiation programme (FLIP). This programme begins with the expression of three genes LEAFY, APETALA 1 and CAULI FLOWER, which specify the floral meristem by activating the floral patterning genes. This is a critical phase in flower development. The flower meristem undergoes a series of developmental change that eventually give rise to the four basic structures of the flower — sepals, petals, stamens and carpels. Each of these structures is derived sequentially from a whorl that develops from the floral meristem.

ANATOMY OF SEPALS AND PETALS

The external structure of sepals and petals may be leaflike. However the internal structure only of green sepals resembles that of foliage leaves and the internal structure of coloured sepals and petals is distinctly different.

In the perianth the vascular system is poorly developed and the veins usually lack sclerenchyma. Usually the epidermal cell walls are thin. In many flowers the anticlinal walls of these cells are folded or undulated and are arranged so that the projections of the cells dove-tail one into the other, a feature that strengthens epidermis. At the base of the petals, as well as along the length of the veins, the anticlinal walls are usually straight even when they are wavy in the other epidermal cells. The outer walls of the epidermal cells commonly have papillae, which make the petals glisten. More papillae are present on the adaxial than the abaxial epidermis and they are not developed at the base of the petals. The epidermis of both sepals and petals may have stomata and trichomes. The stomata on the petals either resemble those on the foliage leaves or are incompletely differentiated. Stomata, if present, are scanty and non functioning. Sometimes intercellular spaces covered by cuticle are formed in the epidermis of petals. The thickness of the cuticle varies in different plants and it may be variously sculptured. Various pigments are found in the epidermal cells of the petals and sepals.

Green sepals contain chloroplasts but rarely show differentiation into palisade and spongy parenchyma. The color of petals, results from pigments in chromoplasts (carotenoids) and in the cell sap (flavonoids mainly anthocyanins) and from various modifying conditions such as, for example, acidity of cell sap.

Crystal containing cells, laticifers, tannin cells and other idioblasts may be present in sepals and petals.

Stamen

The most common type of stamen in angiosperms consists of an anther divided into four microsporangia and born on a thin single veined stalk or filament.

Anther generally contain four microsporangia which are paired in two lobes. The two lobes are seperated by a zone of sterile tissue, termed the connective, through which a vascular bundle passes.

In majority of the Angiosperms anthers are tetrasporangiate. But in Malvaceae, Adoxaceae, Circaestraceae, Epacridaceae and Restionaceae anthers are bisporangiate. In Bixaceae anthers are 8-sporangiate. Both bisporangiate and tetrasporangiate anthers are seen in families Asteraceae, Araceae, Asclepiadaceae, Celastraceae, Lauraceae, Lemnaceae, Moringaceae and Piperaceae. Only one microsporangium per anther has been recorded in *Arceuthobium minutissimum*..

In the young stage, the anther in cross section consists of an oval mass of meristematic cells surrounded by an epidermis. When a four-lobed contour is formed, archesporium gets differentiated hypodermally at the four corners of the anther. The archesporial cells are larger in size, slightly radially elongated and have prominent nuclei. The archesporial cells undergo periclinal divisions forming primary parietal layer and a primary sporogenous layer. The primary parietal layer undergoes periclinal and anticlinal divisions forming the anther wall.

The anther wall, at the pollen mother cell stage comprises of the epidermis, endothecium, middle layer(s) and tapetum. Epidermis is single layered and is the external covering of the sporangium. The epidermis at first keeps pace with the developing internal tissues by anticlinal divisions, but soon, due to the pressure exerted by the developing cells inside the sporangium, its cells become radially flattened and stretched tangentially. The epidermis performs its usual protective function. The layer of cells lying immediately beneath the epidermis is the endothecium. Its cells enlarge and develop fibrous thickenings on their inner tangential walls at maturity. These bands extend radially outward and upwards terminating near the outer tangential wall. The presence of fibrous bands, differential expansion of the outer and inner tangential walls and hygroscopic nature of the endothecial cells help in the dehiscence of anthers at maturity. In majority of angiosperms there is a single middle layer but in some families middle layers are 2-many. The cells of the middle layer ultimately become flattened and crushed at the time of meiotic divisions in the pollen mother cells. The tapetum is the inner most layer of anther wall and surrounds the sporogenous tissue completely. The tapetum is composed of single layer of cells which are full of dense cytoplasm and prominent nuclei. In angiosperms the anther tapetum is of two types, Glandular or Secretary and Amoeboid or Periplasmodial. In glandular type tapetal cells remain *in situ* till the maturation of pollen grains and provide nutrition. In periplasmodial tapetum the walls of the tapetal cells break down, protoplasts enlarge, enter the locule and fuse to form a coenocytic periplasmodium.

The primary sporogenous cells undergo mitotic divisions and form pollen mother cells. The pollen mother cells undergo meiotic divisions resulting in microspore tetrads. The microspores separate from the tetrad and develop into pollen grains.

The filament is relatively simple in structure. The filament consists of parenchyma with well developed vacuoles, small intercellular spaces, often pigments are present in the cell sap. Parenchyma surrounds the vascular bundle which may be amphicribal, The cutinized epidermis may have trichomes, and on both filament and anther stomata may be present. The vascular bundle traverses the entire filament and ends blindly either at the base of the anther or in the connective tissue located between the two anther halves.

The primitive dicotyledonous taxa have stamens which are leaf like (Laminar stamens). A very primitive type of stamen is found in the genus

Degeneria, here no filament, anther or connective can be distinguished as the stamen is broad and leaf like, and has three vascular bundles. The four microsporangia are deeply sunk into the abaxial side of the stamen. The microsporangia are found between the lateral and medium bundles. Similar stamens are found in other Ranalean genera, e.g., *Austrobaileya, Himantandra* and certain genera of the Magnoliaceae. In the Magnoliaceae intermediate stages may be found from broad stamens with three bundles and laminal microsporangia, i.e., microsporangia distant from the margins, as in *Degeneria*, to stamens with marginal microsporangia and distinct filaments and anthers.

Carpel

In primitive angiosperms the carpels appear as folded styleless structures with the stigmatic tissue covering the unsealed margins. In *Degeneria* the carpel is a folded organ of which the margins flare outwards and remain unfused for a long time. This type of carpel is termed conduplicate. Swamy observed on the flared margins and extending into the locule, even beyond the attachment of the ovules, an excessive development of hairs. The cleft between the closely appressed margins is filled with interlocking hairs. On the inner side of the carpel the trichomes are small and papilla-like. According to Swamy all the hairs together form the surface area of the stigma. When pollination takes place the pollen grains become attached to the hairy, outwardly flared margins of the open carpel. Swamy also observed that the germination tubes of pollen grains grow into the carpel locule between these hairs and that they never penetrate the carpel tissue. In the early stage of carpel development the two rows of ovules are very remote from the true margins. The vascular supply to the ovules may be derived from the ventral or dorsal bundles or from both.

In *Drimys* also the carpels are folded in half and they have widened margins with stigma like characteristics. Bailey and Swamy (1951) found that the young carpel has a stalk, and the blade-like portion of the carpel encloses two rows of ovules which are situated on the inner adaxial surface a great distance from margins. When such a mature carpel is cleared and unfolded, it can be seen that the ovules, which are situated between dorsal and ventral bundles, are supplied with traces from both these bundles. Here also the pollen grains reach the hairy margins of the carpels and the pollen tubes penetrate to the ovules by growing between the hairs on the inner surface of the carpels.

A structure somewhat similar to this has been found in the genus *Cananga* of the Annonaceae (Periasamy and Swamy, 1956).

According to Bailey from this type of conduplicate carpel with flared margins, the present carpel has developed by closure and the concentration of the stigmatic margins to the upper part of the carpel only. According to

Bailey, the loss of the marginal portions outside of the ovules has resulted in the impression that the carpel margins are involuted and that the placentae are marginal. During the course of evolution the number of ovules was reduced, and they remained only on the lower portion of the carpel while the upper part underwent differentiation to form style and stigma.

The carpel is classically interpreted as the basic unit of the gynoecium. A flower may have one carpel or more than one. If two or more carpels are present, they may be free from one another (apocarpous gynoecium) or they may be united (syncarpous gynoecium). An old established term used with regard the gynoecium is pistil. It refers to a single carpel in an apocarpous gynoecium (simple pistil) or to an entire syncarpous gynoecium (compound pistil).

As a foliar structure the carpel is described as folded lengthwise in such a way that the adxial surface is enclosed and the margins are more or less completely united. According to the prevalent view, the folding is conduplicate, with margins remaining flat (Fig. 15.1 A). Evolutionary changes in this type of carpel are thought to have led to a reduction of the marginal areas (Fig. 15.1 B). An older concept implies that the folding includes an

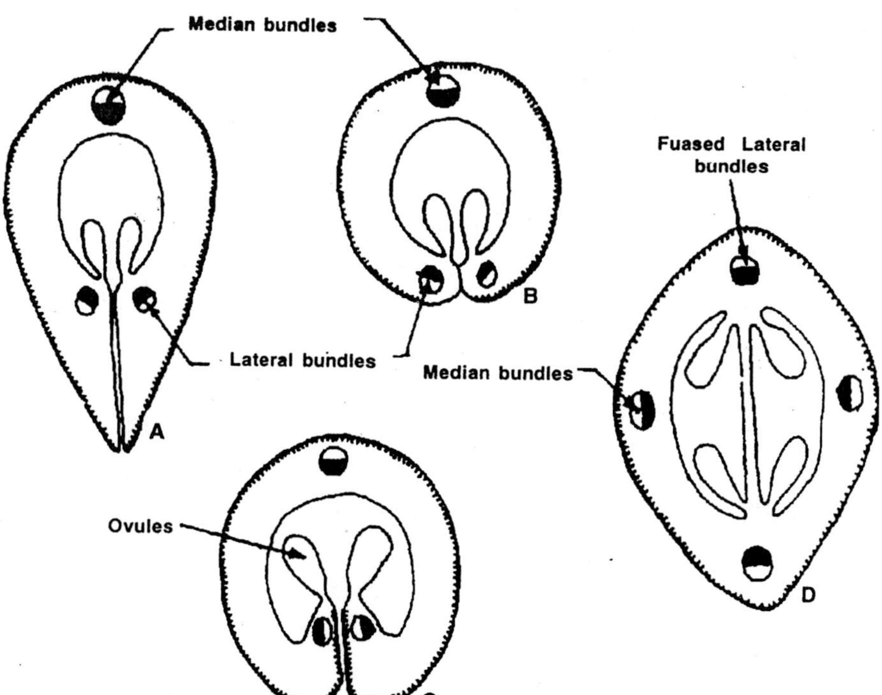

Figure 15.1: Diagrams of cross sections of ovaries composed of one (A–C) and two (D) carpels.

involution of the margins so that the surface is lined with the abaxial surface (Fig. 15.1 C).

The union of carpels in a syncarpous gynoecium follows two basic plans: the carpels are joined either in a folded condition, abaxial surface to abaxial surface (Fig. 15.2 C) or in an unfolded or partially folded condition, margin to margin (Fig. 15.2 D). The first type of arrangement results in a bilocular or multilocular gynoecium (Fig. 15.2 A,C), the second, in a gynoecium with one locule (Fig. 15.2 D). Secondary modifications may bring about variations in the basic plans. One example of such modification is found in the Brassicaceae in which two carpels are united by their margins but the gynoecium is not unilocular because a partition is formed by an outgrowth from the placentae at carpel margins (Fig 15.1 D).

The carpel of an apocarpous gynoecium or the entire syncarpous gynoecium is commonly differentiated into the lower fertile part, the ovary and the upper sterile part, the style. A more or less extensive peripheral part of the style is differentiated as a stigma, usually in the upper region of the style.

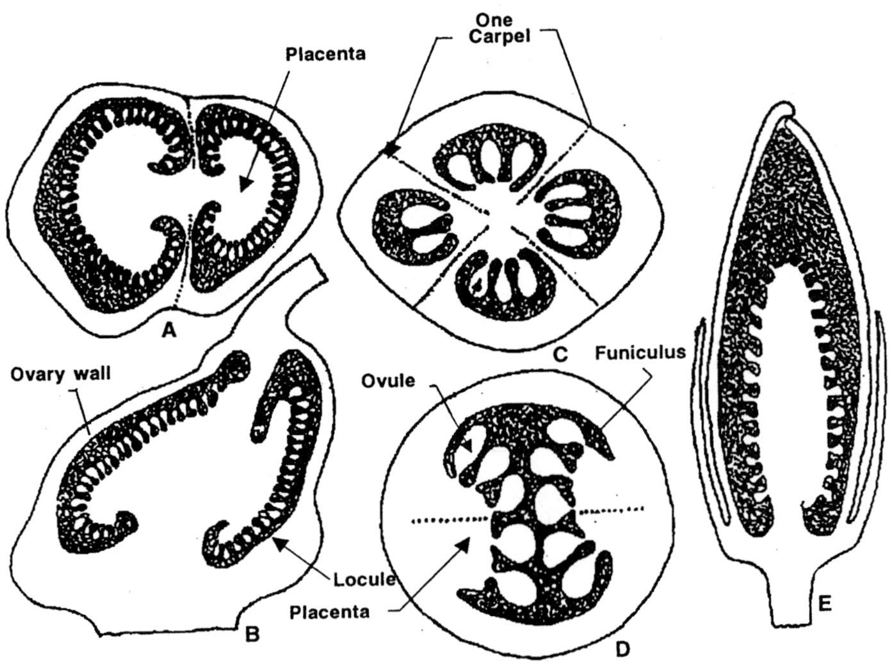

Figure 15.2: Placentation in ovaries in cross (A, C, D) and longitudinal (B, E) sections. (A, B) *Antirrhinum*; (C) *Fuchsia*; (D) *Ribes*; (E) *Dodecatheon*.

Ovary

Structurally ovary consists of ovary wall, the locule (cavity) or locules, and in a multilocular ovary, the partitions. The ovules are borne on the ovary wall located on its inner (adaxial) side. An ovule bearing region constitute the placentum. Each carpel has two placentae. In a given carpel, the placentae occur either close to the margins or at some distance from them. On this basis, one distinguishes between marginal and laminar placentations. The later is considered to be the primitive. A placentum may be conspicuous outgrowth, sometimes almost occluding the lumen of the ovarian cavity.

Structure and Function of the Stigma and Style

The stigma is the receptive surface of the pollen and it consists of specialised cells. The morphology of the stigma and style show great diversity and to some extent depends on the type of pollination. The function of the stigma is to accept the pollen and permit its germination. The stigma also provides nutrients to the pollen and directs the pollen tube growth. Both the pollen coating and stigma coating are involved in recognition and receiving signals.

The morphology of the stigma shows great diversity. Based on the morphology of the stigmatic surface, the amount of secretion and the nature of the surface cells, a general classification is suggested by Heslop-Harrison and Shivanna (1977). They have divided the stigmas into two categories namely — Wet stigmas (with copious fluid secretion), e.g., *Petunia* and Dry stigmas (limited secretions). In dry stigmas, the receptive cells are dispersed on multiseriate branches (*Zea, Sorghum, Hordeum* etc.) or concentrated in distinct ridges or zones. However, in wet stigmas, the receptive surface may have papillae (Liliaceae, Orchidaceae etc.) or non-papillate surface (*Dracaena, Alpinia* etc.) Trinucleate pollen is mostly associated with the dry stigmas, whereas binucleate pollen is seen in both cases.

Structural differences have been noted in the stigmatic papillae of floral morphs in many heterostylous plants. Ghosh and Shivanna reported that in *Linum grandiflorum* the stigma of the pin morph (long styled) is of the 'dry' type and smooth and the papilla is covered by a distinct cuticle pellicle layer. The stigma of the thrum morph (short styled), on the other hand, is of the wet type, warty and uneven and the cuticle-pellicle layer is disrupted.

Dumas (1978) classified the stigmas based on the function and the nature of secreted products. Stigmas have an abundant secretion of lipophilic or polysaccharide substances or in some cases, less abundant secretions with polyphenols, tannins, flavonoids etc. Some others have no liquid secretion at all, but stigmas are covered with waxes or proteins. This classification can also be combined with the stylar structure. Hollow styles of monocots produce mainly polysaccharides whereas solid styles of dicots, produce a lipophytic exudate. Some families like Orchidaceae, Solanaceae, Scrophulariaceae excrete lipopolysaccharide substances.

Nature of the Stigma Covering: The receptive surface is formed by the surface cells which are glandular in nature. This is followed by a transition zone of stigmatoid tissue without glandular cells. The stigmatoid tissue is also known as the neck of the transmitting tissue.

The dry stigma has a cuticle covered with a pellicle. The cuticle has some discontinuties for the water. Both the pellicle and the surface of the stigma remain hydrated. The pellicle is an extracellular product originated from the proteins of the stigmatic cells.

The exudate of wet stigmas mainly consists of amino acids, lipids, antioxidants, proteins and alkaloids. During pollination, the exudate serves as a nutritive for birds, bees, beetles, bats, moths etc. The exudate also had boron and serve as a nutrient for pollen. The exudate of wet stigmas varies in quality and composition and can be considered as a complex fluid. In *Trifolium*, it contains glucose, sucrose and traces of galactose, arabinose, some proteins and esterases. The exudate functions in the hydration of the pollen grain, stimulation or inhibition in pollen germination as well as in nutrition.

Fresh exudate of *Phaseolus* contains lipids, proteins, amino acids, sugars, namely glucose, reducing acids and phenols. Alkaloids are absent. Sugars stimulate pollen germination; lipids prevent desiccation and also attract insects. Proteins function in recognition and the amino acids as nutrient for insects. Phenols function as protection against chewing insects and pathogens. Thus, the nature of the stigma covering is very complex and it is not possible to make any generalizations.

Structure and Function of the Style: The morphology and anatomy of the style are variable. Three different types of styles are seen in angiosperms namely open, solid and half closed.

Open type (Hollow Style) : This type is more common in monocotyledons, e.g., *Lilium, Amaryllis, Gladiolus.* In the hollow style there may be one or more canals; the number of canals normally corresponds to the number of carpels. The stylar canal is lined with canal cells. Pollen tubes creep on the surface of these canal cells. The canal cells are secretory in nature. The canal cells show dense cytoplasmic contents and are rich in cell organelles. The secretion from the canal cells contains protein, carbohydrates, lipids and shows activity for esterases and acid phosphatases. It has been suggested that in *Lilium* the parenchyma cells underneath the canal cells synthesize some mucilaginous substances and transport them to the canal cells, which, in turn, secrete them into stylar canal. After pollination the canal is filled with secretions which serves as nutrition for the growing pollen tubes.

In the open style, canal cells provide elements for pollen tube growth and functionally similar to the intercellular substances of the solid style.

Solid type: Mostly seen in dicots, e.g. *Datura, Gossypium*. It has a solid core of transmitting tissue. The transmitting tissue of the solid style has either loosely arranged tissue or it may be more compact. This depends on the amount and quality of the intracellular substance. In general, the cells of the transmitting tissue show a thick cell wall with a small lumen or a thin cell wall with large amount of intracellular substance. The cells of the transmitting tissue contain numerous mitochondria, plastids, Rough endoplasmic reticulum, dictyosomes and ribosomes. Along the transmitting tissue, carbohydrates and isoenzymes such as peroxidase and acid phosphatase are found. The function of the intercellular substance is aiding in the pollen tube growth. In *Oryza sativa* the solid style is devoid of any transmitting tissue.

Half closed type: It is present in a few taxa in which the transmitting tissue can be limited to one side of the stylar canal. It is seen in family Cactaceae.

Vascular System

The position of placentae in an ovary is related to the method of union of carpels. The placentae of marginally joined carpels are said to be inserted on the ovary wall and the placentation is called parietal. If the carpels are joined in a folded condition, the ovary is bilocular or multilocular and the placentae occur in the centre of the ovary where the carpel margins meet. This kind of placentation is called axile. Various modifications of these two basic placentations occur in different taxa. The partitions in a multilocular ovary may disappear in terms of either phylogeny or ontogeny so that a free central placentation results. The placenta in a unilocular ovary may occur at the base and the placentation is then called basal.

In relatively unspecialized flowers the vascular organization is readily interpreted as being comparable to that of vegetative shoot. The number of traces of the different floral parts varies in the same flower. The usual pattern is as follows. Each sepal has as many traces as the foliage leaf of the same plant. Each petal in a dicotyledon has one trace; each tepal in a Monocotyledon, one to many. Within the sepal and petal the vascular bundles form more or less complex systems resembling those in foliage leaves. The specialized type of stamen has one trace, which is continued as a single bundle in the filament and anther. The carpel has three traces and their prolongations within the carpel may have branches. The ovules are supplied by branches from carpellary bundles, usually the two lateral ones. Carpellary vascular bundles are continuous through the style. The vascular bundles of floral parts may be fused when these are united.

In some epigynous flowers certain bundles show an inverted orientation of xylem and phloem. Such arrangement is interpreted as an indication that the ovary is embedded in receptacular (axial) tissue, whereas absence of inverted bundles is regarded as evidence that the extracarpellary tissue is appendicular in nature (floral tube).

CHAPTER XVI

THE FRUIT

A fruit is a developed and ripened ovary or ovaries together, with adjacent floral organs and other plant parts. Such organs may be tepals (*Morus*), the receptacle (*Fragaria*), bracts (*Ananas*), the floral tube, which is formed by the floral organs together with receptable (*Pyrus malus*) or the enlarged axis of the inflorescence (*Ficus*).

Fruit generally develops after fertilization, but in many plants fruits may develop without fertilization and seed development. This phenomenon is termed parthenocarpy.

The body of the fruit developed from the ovary wall, which surrounds and encloses the seed, is known as pericarp. When this is not homogenous histologically, and distinct outer, median and inner parts are evident, these parts are known as exocarp, mesocarp, and endocarp. In many fruits the pericarp is more or less nearly homogeneous and not separable into parts, in other fruits only exocarp and endocarp are evident. Strictly speaking, the "Pericarp" refers only to the modified ovary wall, but in general usage it is applied to the outer tissues of fruits regardless of their morphological nature.

On the basis of fruit wall histology fruits may be distinguished into Dry fruits and Fleshy fruits.

DRY FRUIT

A dehiscent fruit may develop from a single carpel (Follicle, Legume) or from more than one carpel (Capsule, Siliqua).

The method of dehiscence is highly variable. If the ovary is derived from a single carpel, the break may occur longitudinally through (1) the suture resulting from the union of carpel margins; (2) the back of the carpel; (3) the suture and the back simultaneously (Fig. 16.1 A–C). In syncarpous fruits with parietal placentation, dehiscence may occur through the suture between the carpels or through the backs of carpels (Fig. 16.1 D–F). In syncarpous fruits with axile placentation (Fig. 16.1 G), the separation along the line of union of contiguous carpels, that is, through the septae (septicidal dehiscence), may be combined with breaking away from the central column of tissue (Fig. 16.1 H). Dehiscence through the backs of the carpels opens the individual locules (loculicidal dehiscence, Fig. 16.1 I). In the fruit types mentioned above the longitudinal dehiscence may also occur without reference to the suture or the median vein. In some fruits, dehiscence occurs through a horizontal circular region involving all carpels (circumscissile dehiscence) or through pores (poricidal dehiscence).

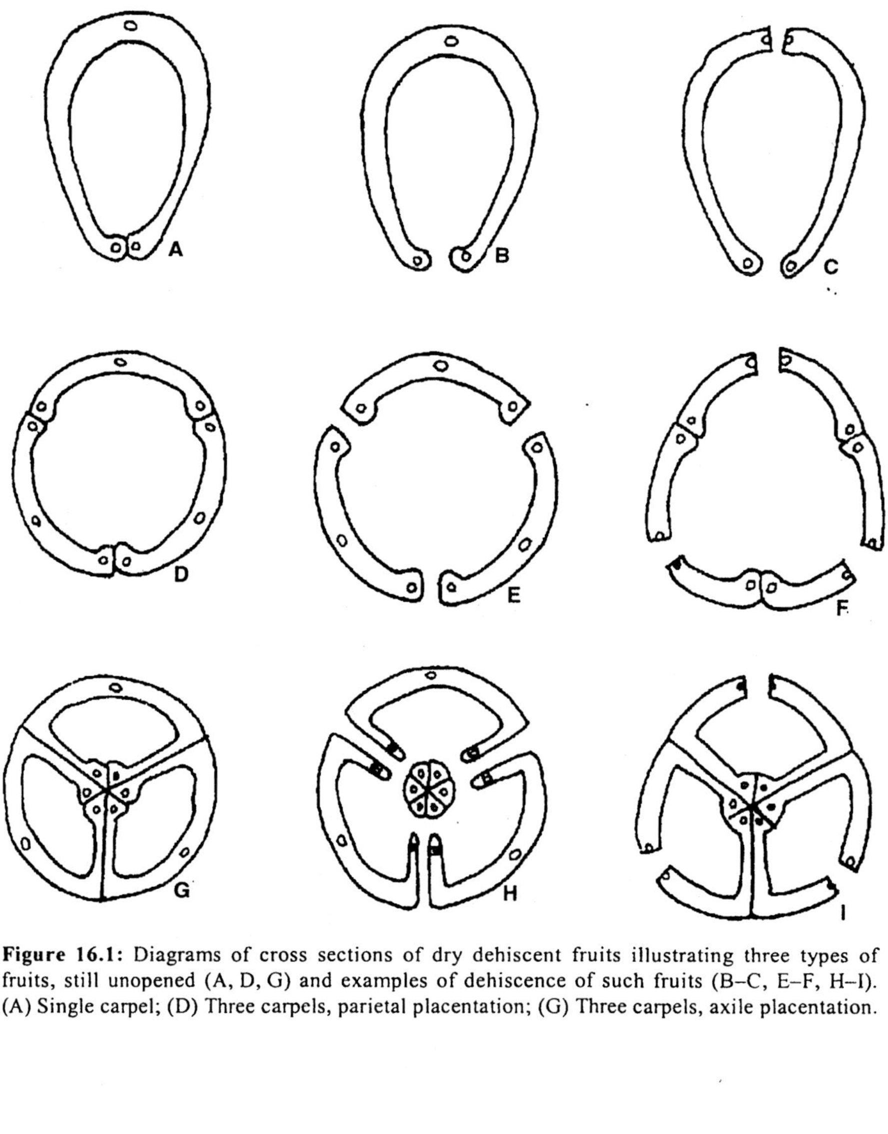

Figure 16.1: Diagrams of cross sections of dry dehiscent fruits illustrating three types of fruits, still unopened (A, D, G) and examples of dehiscence of such fruits (B–C, E–F, H–I). (A) Single carpel; (D) Three carpels, parietal placentation; (G) Three carpels, axile placentation.

Legume

Legume, found in many Fabaceae, is a well known example of a dry dehiscent fruit. The legume is derived from a superior ovary which embodies a single carpel. It dehisces along the suture of carpel margins and along the median vein (Fig. 16.1 C). As seen in *Glycine max* the exocarp is composed of outer epidermis and hypodermis, both with thickened cell walls. The mesocarp is composed of several layers of parenchyma while endocarp is composed of sclerenchyma on the inside of which is usually a thin-walled epidermis. The vascular bundles are situated in the parenchyma of the mesocarp. The cells in the hypodermis and the sclerenchyma of the endocarp are elongated but the long axes of the two kinds of cells are oriented in opposite planes and the developing stresses promote the opening of the dehydrated mature fruit. Another feature found to be associated with the differential shrinkage of the pericarp in some legumes is the dissimilar orientation of cell wall microfibrils in different layers of the sclerenchyma, in low-pitched helices in some, in steeply pitched helices in others. Tensions developing during dehydration in some legumes are so great that the separated valves of the pod become twisted.

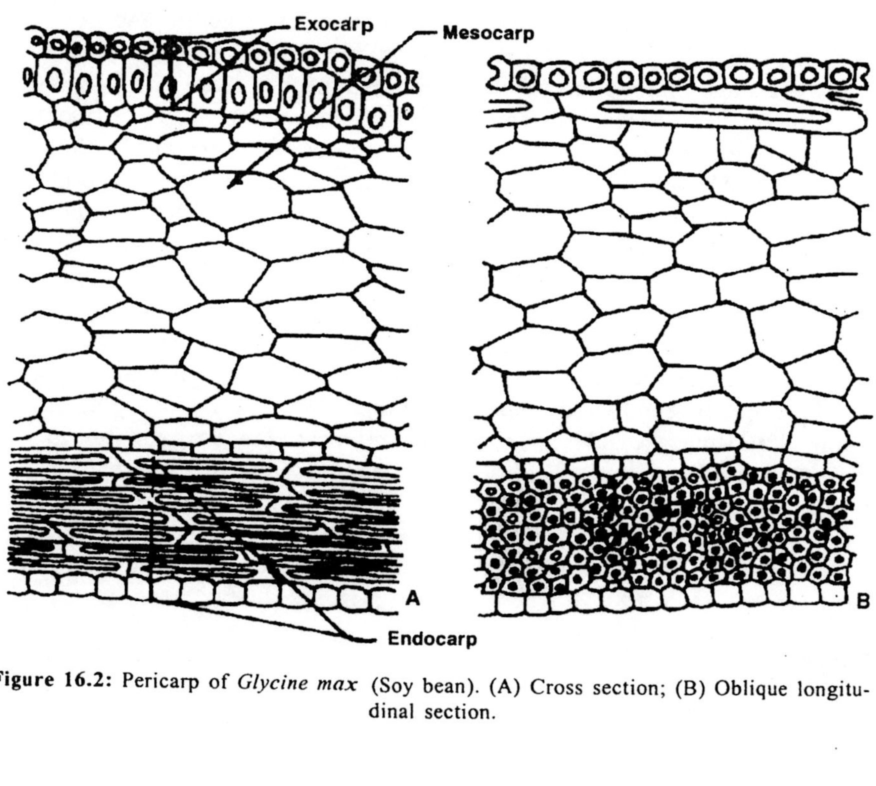

Figure 16.2: Pericarp of *Glycine max* (Soy bean). (A) Cross section; (B) Oblique longitudinal section.

In *Phaseolus* the inner epidermis undergoes numerous periclinal divisions and becomes a massive layer of usually non-photosynthetic inner parenchyma.

Capsule

The outer part of the pericarp consists of rigid, highly lignified cells, the inner part of parenchyma cells. The inner epidermal cells may also be thick-walled. The dehiscence of capsule is brought out about by the anisotropic swelling of the cell walls. The walls that bring about the dehiscence are mainly the very thick walls of the outer or inner epidermis, and it is they that determine the direction of the bending of the teeth or valves.

INDEHISCENT FRUIT

An indehiscent fruit usually originates from an ovary in which only one seed develops, though more than one ovule may be present. The pericarp of an indehiscent fruit often resemble seed coat in structure. The actual seed coat in such fruits may become obliterated to a considerable degree (achene in Asteraceae) or fused with the pericarp (Caryopsis of the Poaceae).

Cypsela: Pullaiah (1979) described the anatomical structure and development of pericarp of *Adenostemma lavenia,* a member of Asteraceae. At fully organised stage of the embryo sac, the integument consists of 6–10 layers in addition to outer epidermis and integumentary tapetum. After fertilization the cells around the integumentary tapetum become enlarged and depleted in appearance. This degeneration extends into the peripheral region and all the integumentary layers except the epidermis have become crushed, by the time a well-developed embryo has formed. Meanwhile the integumentary tapetum also becomes crushed. Thus, in the fully mature seed it is only the epidermis that forms seed coat. The ovary wall at the organized embryo sac stage, consists of an epidermis, a hypodermis and an inner layer of parenchymatous cells traversed by vascular elements. After fertilization schizogenous cavities develop between hypodermis and the inner layers of the ovary. A brown resinous substance exudes from the hypodermis and accumulates in the schizogenous cavities. At maturity this resinous substance hardens into a dark mass. Meanwhile, the inner layers of the vary wall, except for a few layers enlarge and then become obliterated. In the mature achene the pericarp thus consists of epidermis, the hypodermis, a hardened dark layer and a few inner layers.

Caryopsis

In grasses the pericarp and the remains of the integuments of the single seed of the caryopsis are completely fused. The covering layers of wheat caryopsis are composed of the pericarp and the remains of the seed coat. Five layers can be distinguished in the pericarp: the outer epidermis, the hypodermis, a zone of thin walled cells, cross cells and tube cells. The outer epidermis and the hypodermis together form the exocarp. The cells of the exocarp are elongated in a direction parallel to the longitudinal axis of the caryopsis, and they become compressed and their walls thicken considerably. The cross cells are found below the parenchymatous layer and they have thick walls with pits which are elongated transversely to the cell. The longitudinal axis of these cells is at right angles to that of the exocarp cells. The tube cells constitute the inner epidermis of the pericarp and they occur on the inside of the cross cells. The longitudinal axis of the tube cells is parallel to that of the exocarp cells. The seed coat, which is united with the pericarp, is crushed in the mature caryopsis. In the development of the seed coat, the outer integument disintegrates, and the inner becomes altered and compressed. The compressed inner integument contains pigment. With in the seed coat lies the endosperm which forms about 83 percent of the fruit.

In *Zea* the outer pericarp consists of cells with thick pitted walls and is much compressed. The integuments disintegrated completely.

Cremocarp

The fruit of *Apium graveolens* is described here as an example of a cremocarp. Each mericarp has five ribs in which the vascular bundles, accompanied by sclerenchyma are situated (Fig. 16.3 A–C). Between the ribs one to three oil ducts may be present (Fig. 16.3 B–D). The exocarp cells (i.e., the outer epidermis) are small and isodiametric. In surface view they appear lobed. The mesocarp is parenchymatous and its cells have small papillae and finely striated walls. The oil ducts in the mesocarp are surrounded by polyhedral cells which become brown as the fruit ripens. The slightly elongated cells of the inner most layer of the mesocarp are wider than those of the endocarp. The endocarp (the inner epidermis) consists of narrow cells of which the longitudinal axis is parallel to the transverse axis of the mesocarp (Fig. 16.3 C, D) or their arrangement is mosaic. Inside the pericarp is the thin seed coat which surrounds the endosperm in which the embryo is completely enclosed. The maturing mericarps, each of which develops from a single carpel together with a small amount of tissue outside it (inferior ovary), separate one from the other along the area of adnation between them, but they remain attached to the branched carpophore (Fig. 16.3 E).

FLESHY FRUIT

Hesperidium

The *Citrus* fruit, called hesperidium develops from a superior ovary and has about ten carpels (Fig. 16.4 A, B). The placentation is axile. The exocarp or

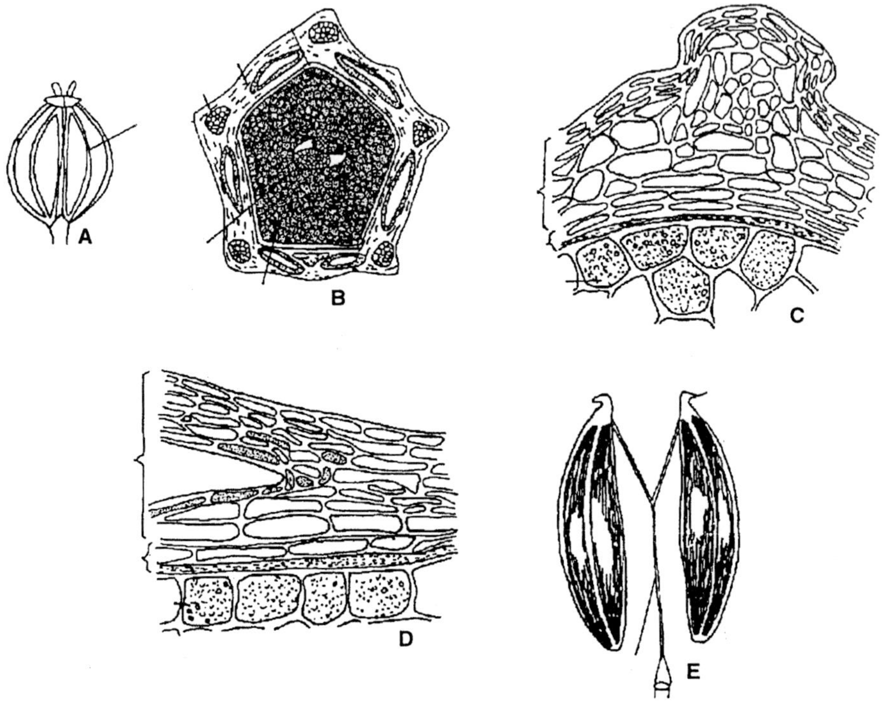

Figure 16.3: Fruit of *Apium graveolens*. (A) Entire cremocarp; (B) Cross section of a single mericarp; (C) Portion of a cross section of a mericarp showing the structure of a rib; (D) As in C, in the region of oil cavity; (E) A mature cremocarp with separated mericarps.

flavedo (yellow tissue) consists of a cuticle-covered outer epidermis and compact subepidermal parenchyma with oil glands and crystal-containing cells. The mesocarp or albedo (white tissue), consists of parenchyma with large intercellular spaces and cells with armlike extensions (aerenchyma). A vascular network occurs in this tissue. The endocarp is composed of the inner epidermis and a few layers of compact parenchyma. The partitions between the locules are extensions of endocarp and mesocarp tissues. The endocarp produces the juice sacs that completely fill the carpel locules (Fig. 16.4 D). The stalked, spindle-shaped juice vesicles, which fill the locules when the fruit ripens, develop from the cells of the inner epidermis and subepidermal layers (Fig. 16.4 C). Each juice vesicle is covered externally by a layer of elongated epidermal cells which enclose very large, extremely thin-walled juice cells. Development of juice sacs begins at the time when the flower opens and continues until the style withers. The chloroplasts, responsible for the green color of immature fruits, occur in the parenchyma of the exocarp. They become changed into chromoplasts as the fruit matures.

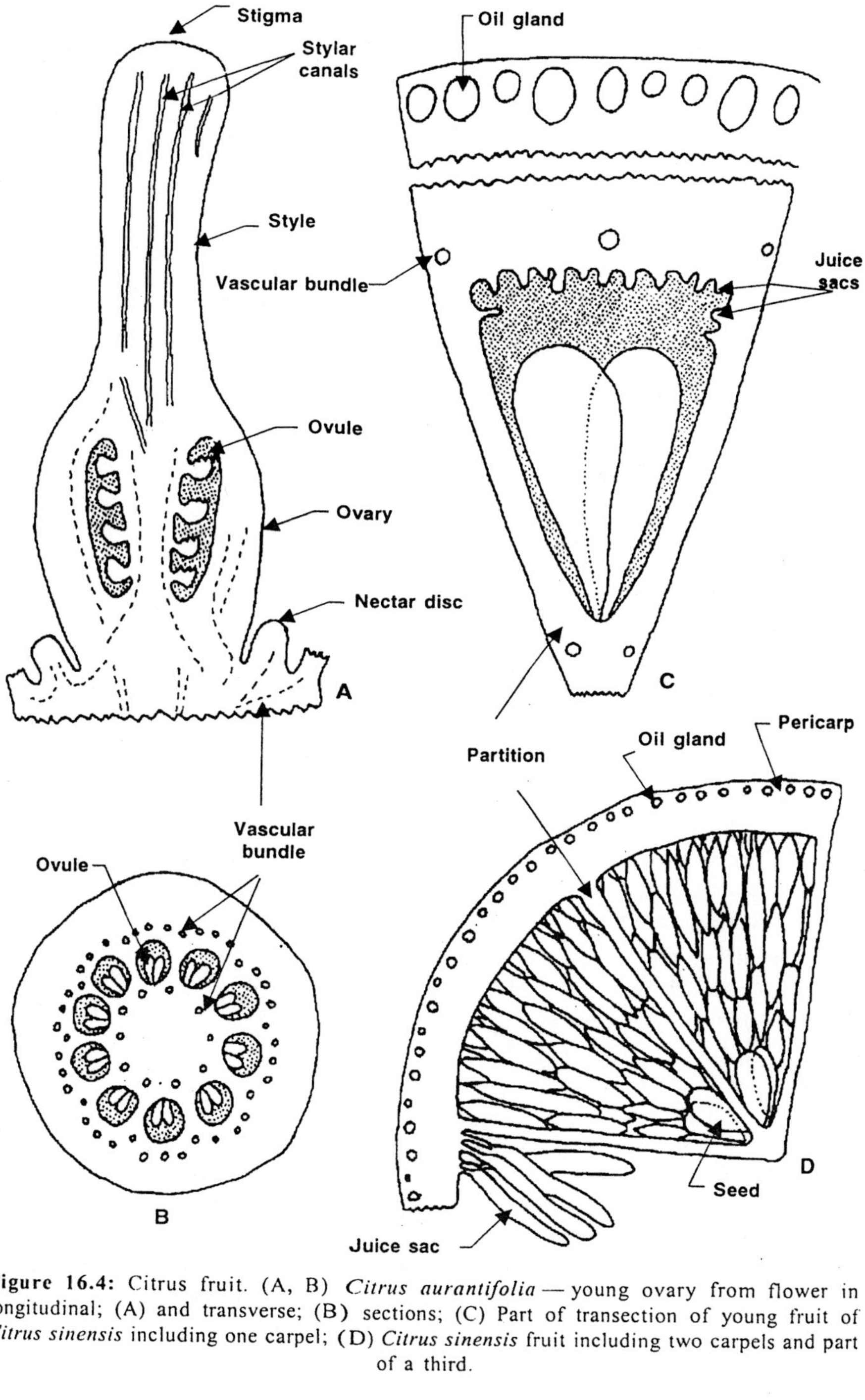

Figure 16.4: Citrus fruit. (A, B) *Citrus aurantifolia* — young ovary from flower in longitudinal; (A) and transverse; (B) sections; (C) Part of transection of young fruit of *Citrus sinensis* including one carpel; (D) *Citrus sinensis* fruit including two carpels and part of a third.

Berry

The fruit of tomato (*Lycopersicon esculentum*) consists of a pericarp and a placental tissue on which the seeds are borne. The exocarp consists of an epidermis and three or four layers of collenchyma cells. The epidermal cells are polyhedral and are covered by a thin cuticle. The glandular and other hairs that are usually present on the young fruit are shed as it matures. There are no stomata on the epidermis of the fruit. The mesocarp consists of a thick layer of large, thin-walled cells which enclose many intercellular spaces. In the early stages of development, soon after pollination, the number of layers in the mesocarp increases rapidly but the main increase in thickness of the pericarp results from an enormous increase in cell volume in later stages of development.

The change in the colour of tomato fruit during ripening occurs through transformation of chloroplasts into chromoplasts. Two kinds of chromoplasts develop. One, located in the gelatinous tissue, contains globular inclusions with β-carotene. The other, located in the pericarp, includes voluminous paracrystalline sheets of lycopene in addition to the globules with β-carotene. Towards maturity the placental tissue becomes degraded and assumes a gelatinous consistency.

Banana

Musa paradisiaca var. *sapientum* is another good example of berry. It's wall consists of (i) outer epidermis covered by cuticle, (ii) multilayered parenchymatous hypodermis, (iii) a broad parenchymatous region in which vascular bundles are found scattered, (iv) another region of parenchyma with large number of well developed air spaces, (v) a zone in which vascular bundles are surrounded by laticiferous cells, and (vi) a five to seven layered parenchymatous region and an inner epidermis. The banana fruit has tannin containing cells in the pulp and skin which impart astringent taste to unripe fruits. In ripened fruits, the amount of tannin is gradually reduced.

Apple

The fruit of *Pyrus malus,* the pome, arises from an inferior ovary. The fruit here mainly develops from the hypanthium, i.e., from the basal portions of the perianth and stamens and which are fused and adnated to carpels. The pericarp may be divided into a relatively thin exocarp composed of the epidermis and the subepidermal collenchyma, a fleshy mesocarp and a stony endocarp. The epidermis is covered by a cuticle which increases in thickness as the fruit develops. Stomata and trichomes occur in young fruits. Lenticels later replace the stomata. The subepidermal tissue is compact and has thickened walls. It forms, together with epidermis, the relatively tough "skin" of the fruit. Anthocyanins occur in the skin of the red varieties of apple. Chloroplasts which change into chromoplasts with carotenoids during the ripening process, are also located in the peripheral tissues of the fruit.

The outer parenchyma of the fruit develops from the hypanthium and it contains ten vascular bundles — five belonging to the sepals and five belonging to the petals. These vascular bundle are branched and the branches penetrate into the parenchyma where they form a network. The bulk of the extra carpellary tissue is parenchyma permeated with intercellular spaces.

The endocarp is cartilaginous and it lines the locules. It consists of elongated sclereids with very thick walls which almost completely obliterate the cell lumen.

In *Pyrus communis* (pear) groups of brachysclereids develop in the parenchyma. The succulent sugar-containing cells are arranged around such groups of brachysclereids in a radial pattern.

Structural changes occur in the walls of the parenchyma cells of apple and pear fruits during the advanced stages of ripening. The middle lamella disintegrates almost completely and the adjoining outer region of the cell wall also undergoes partial dissolution of its cellulosic fibrils become dispersed.

CHAPTER XVII

THE SEED

Ovule after fertilization develops into seed. In the mature seed the following parts can be distinguished: the seed coat, commonly called testa, which develops from one or two integuments, endosperm, which may be present in a large or small amount and the embryo. In some seeds the endosperm is completely absorbed by the developing embryo and no trace of it is left at maturity, such seeds are called as Exalbuminous seeds. In some plants like centrospermae the nucellus persists in mature seed and is called perisperm. The micropyle may remain visible as an occluded pore or may be obliterated. The funiculus, whole or in part, abscises from the ovule and leaves a scar, the hilum, water can penetrate with relative ease-through hilum. In anatropous ovules the part of the funiculus that is adnate to the ovule remains recognizable as longitudinal ridge; the raphe, on one side of the seed.

In his comprehensive seed classification based on internal structure Martin (1946) selected the embryo as the principal diagnostic feature and used its shape, size and position to divide seeds into 12 types grouped into 3 divisions.

SEED COAT

Seed coat develops either from one integument or from both the integuments. The seeds of angiosperms have mostly dry seed coats but some have fleshy appendages such as elaiosome, or juicy layers such as the edible outer epidermis of the seed coat in pomegranate (*Punica granatum*).

All parts of one or both integuments may take part in the formation of the seed coat e.g., *Viola tricolor*. However, in most seeds much of the integumental tissue is destroyed and absorbed by the other developing tissues of the seed, and then the seed coat develops only from the remaining parts of the integuments. The parts that are destroyed are usually the innermost or intermediate middle layers of the integument.

In certain seeds, especially those of indehiscent fruits, only the two or three outermost layers of the integuments persist. In the mature seed of the Apiaceae only the outer unspecialized epidermis of the single integument remains. In the seeds of certain genera of the Asteraceae, e.g., *Lactuca*, the integument is represented only by a thin layer of obliterated cells which persists beneath the fruit wall. The innermost layers of the fruit wall also disintegrate. In the seeds of certain monocotyledons, e.g., *Zea*, the integuments are completely destroyed. In seeds developing from ovules with two integuments, the two integuments may be present in the testa or only two or three of the outermost layers of the outer integument may constitute the major portion of the seed coat or the outer may be the better developed and be specially adapted for protection while the inner is non-specialized. The former type is characteristic of the Malvaceae, Violaceae, Hypericaceae and Tiliaceae and the latter of the Brassicaceae, Papaveraceae, Berberidaceae and certain genera of the Liliaceae, Iridaceae and Araceae. In the Onagraceae, Lythraceae, Aristolochiaceae, and others, protective layers develop from both integuments and the entire nucellus or its outermost layers only also take part in the formation of the seed coat. In most of the Leguminosae, Ranunculaceae and certain genera of the Liliaceae and Amaryllidaceae, the inner integument and the nucellus are completely obliterated. In the ovules with single integument only a few of the outermost layers and the inner epidermis persists, e.g., in the Plantaginaceae and Polemoniaceae.

Seed coats of pulses are impervious to water and gases due to deposition of thick layer of cuticle except in *Cajanus cajan* where it is very thin and smooth. The cuticle is smooth surfaced in *Cajanus cajan, Vicia faba, Dolichos biflorus* and *Phseolus aconitifolius*, rough with dentate structures in *Lens culinaris* and *Lathyrus sativus* and papillose in *Phaseolus aureus*. In pulses seeds cuticle is followed by a layer of characteristic malpighian cells, called palisade layer. It consists of closely packed palisade like cells, that elongate at right angles to the surface and differentiate macrosclereids with unevenly thickened walls. Two palisade layers occur in the hilum region. The outer of these is derived from the funiculus. The cells of the subepidermal layer differentiate into the so-called columnar cells, also termed pillar cells, hour glass cells, osteosclereids or lagenosclereids, depending on the the distribution of wall thickenings and shape of cells. The deeper tissue is a lacunate parenchyma with large tangentially elongated cells in the outer part and smaller much branched cells in the inner part.

The palisade layer attracts much attention because its structure in certain hard legume seeds is assumed to be causally connected with their high degree of impermeability. The so-called light line of the palisade cells is thought to be particularly impermeable region. The light-line effect results from the intense refraction in a restricted region in epidermal walls. In sections of seeds this region is oriented tangentially above the middle of a cell or close to its outer wall. The cell wall in the light-line region appears to be especially compact.

In members of Brassicaceae, the seed coat is partly derived from outer integument and partly from inner integument of the ovule. The bitegmic ovules have rather thick integuments. The outer has two to five cell layers, the inner up to ten. In many species, the epidermal cells of the outer integument become almost filled with mucilaginous material which appears in layers (Fig. 17.1 A, B). The mucilage, which consists of pectin and cellulose, swells when it comes in contact with water and, in some species, bursts the outer wall of the cell and forms a gelatinous film over the surface of the seed. If subepidermal parenchyma is present in the outer integument, it develops thickened walls (Fig. 17.1 B) or becomes crushed (Fig. 17.1 A). The inner epidermis of the outer integument, the palisade layer, is the strongest layer in most species, for its cell develop lignified thickenings on the radial and inner tangential walls (Fig. 17.1 A, B). The inner integument dies and is compressed. In some species the inner epidermis of this integument becomes the pigment layer.

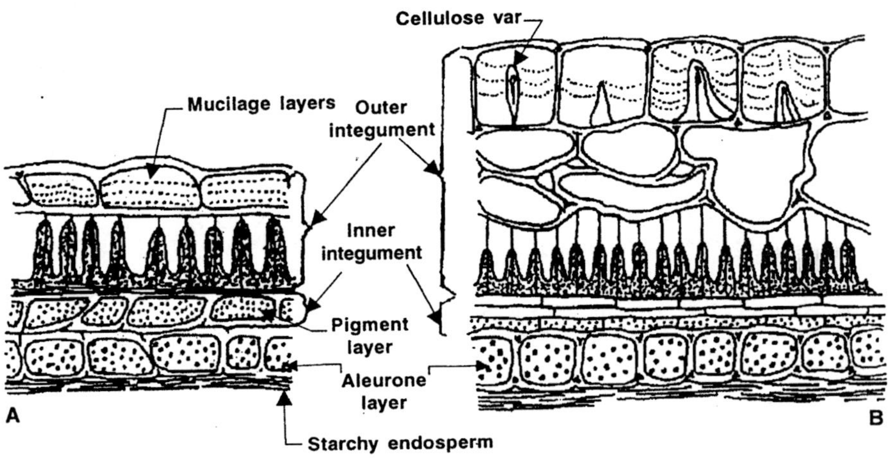

Figure 17.1: Cross sections of seed coat and aleurone layer of (A) *Brassica*; (B) *Sinapsis*.

Corner (1976) in his book "*The seeds of Dicotyledones*" introduced an original classification of the seed-coat constituents. The portion of the seed coat produced by the outer integument is termed *testa* and that produced by the inner integument *tegmen*. Seeds with characteristic testa are called testal and those with characteristic tegmen tegmic. According to which tissue in the seed coat becomes sclerified, Corner suggested the following terms. Exotestal and exotegmic seeds depending on whether the mechanical layer develops from the outer epidermis in the *testa* or *tegmen* respectively, mesotestal and mesotegmic seeds in which the mechanical layer develops in the middle layer of one of the integuments, and endotestal and endotegmic seeds in which the mechanical layer develops from the inner epidermis of the appropriate integument.

In general, the epidermal cells of seed coat are thick walled and remain filled with colouring matter. But there are certain variations, e.g., in *Gossypium*, the epidermal cells elongate into hairs (the cotton fibres) and in *Linum*, the epidermal walls are highly hygroscopic and become mucilaginous on contact with moisture. The epidermis of seed coat of members of Cucurbitaceae consists of highly elongated cells with characteristic presence of branching ribs.

CHAPTER XVIII

DORMANCY

BUD DORMANCY

The growth of deciduous perennial plants typical of temperate regions is cyclical — such plants normally undergo a cessation of active growth on an annual basis. Buds formed in the latter part of the summer or early fall over winter in a dormant condition and do not normally renew their growth until more favorable temperature conditions return in the spring. Bud dormancy is thus a defence mechanism, helping to ensure that the plant is able to survive the adverse conditions of winter. To ensure survival, dormancy mechanisms must be in place before the arrival of unfavorable conditions. This means that the plant must be able to anticipate climate change. Mechanisms must also be in place to ensure that the buds do not break dormancy until such time that conditions in the environment are appropriate to sustain normal growth and development. Premature breaking of buds during an unseasonally warm period in the winter, for example, could have serious consequences. In other words, dormancy is more than simply a period of inactivity. Dormancy is a precisely regulated phenomenon, cued by factors in the environment and maintained, and ultimately broken, by specific metabolic changes in the organism.

The term dormancy is generally used to indicate tissues such as buds, seeds, tubers, and corms that fail to grow even though they are provided with adequate moisture and oxygenatean appropriate (i.e., physiological) temperature.

The onset of bud dormancy is coincident with leaf fall, decreased cambial activity, and increased cold hardiness. Bud dormancy appears to be a typical short day response, initiated by the short days of late summer. In most cases, phytochrome is involved and the leaf perceives the photoperiodic stimulus. However, in at least one case, that of birch (*Betula pubescens*), the response appears to be localized in the apical bud.

A bud is actually a meristem in which the internodes have failed to elongate and leaves have not enlarged. The whole is enclosed in a set of modified leaves, called bud scales. Bud scales serve a protective function, insulating the bud and preventing desiccation. Bud scales also contribute to reduced respiration since their removal usually results in a significant increase in respiratory rate.

Temperature near or just above freezing appear to be most effective at breaking dormancy. The amount of chilling required varies with species, cultivar, and even location of the buds on the trees. Species such as apple (*Malus pumila*), pear (*Pyrus communis*) and cherry (*Prunus* sps.) require approximately 7 to 9 weeks of exposure to temperatures below 7°C in order to overcome dormancy. Others may require up to 22 weeks (*Prunus americana*). Persimmons (*Diospyros kaki*) require only 4 days of low temperature and so can be grown successfully much further south than other fruit trees.

SEED DORMANCY

Many seeds fail to germinate after falling from the plant even if favourable conditions of moisture, temperature and aeration are available. Seed dormancy can be defined as the failure of mature and viable seeds to germinate under favourable conditions for germination. Dormancy occurs because some property of the seed prevents germination. In some dormant seeds morphological changes must take place before commencement of germination. For others, parts of seed must undergo physiological changes before germination can occur. Under natural conditions necessary changes take place gradually under varying combinations of aeration, moisture, temperature and light.

Types of Dormancy

1. *Seed-coat dormancy (External dormancy)*: In many plants the cause of dormancy is the presence of hard seed-coat, which may offer mechanical resistance to embryo growth (Malvaceae, Fabaceae), or it may be impenetrable to water (*Melilotus, Trigonella*) or oxygen (*Xanthium*). In nature such seeds are rendered germinable by the weakening or breaking-down of the seed-coat by microbial attacks, mechanical abrasions, passage through the digestive tract of animals, or exposure to alternating low and high temperatures. Some of the treatments used to break seed-coat dormancy artificially are (a) scarification, which means scratching the seeds with abrasives (e.g., *Cassia, Melilotus*), (b) soaking the seeds in concentrated sulphuric acid for 15–16 minutes (e.g., *Cercopus, Gossypium, Pistacia*), and (c) washing the seeds with alcohol (e.g., *Cassia, Parkinsonia*).
2. *Embryo dormancy (Internal dormancy)*: The most common type of seed dormancy is one in which the embryo requires certain period of time

after harvesting to be able to germinate. During this period, which is called after-ripening period, the embryo may undergo structural maturation or physiological maturation. The most likely cause of embryo dormancy is the presence of germination inhibitors in the embryonic axis or in the food storage tissues or the seed. For germination to occur, these inhibitors must be metabolically inactivated or germination promoting substances must overcome their effect. There is considerable evidence that ABA is involved in regulating the onset of dormancy and in maintaining the dormant state. Germination inhibitors have been isolated and identified in a number of woody plant seeds with ABA the most common inhibitor. ABA accumulation in developing seeds is low during the early stages and is greatest during mid-development. So prevention of germination during development may be due the endogenous ABA content of the seed, e.g., *Acer platanoides, Malus pumila, Quercus rubra.*

3. *Morphological Dormancy*: This refers to an under-developed undifferentiated embryo. Morphological dormancy results from the embryo not being completely developed when seeds are shed. At shedding, the seeds in some members of the Ranunculaceae contain an undifferentiated embryo. During the after-ripening period the embryo undergoes further development and acquires the mature structure. In *Fraxinus excelsior*, a seed when shed contains a fully differentiated embryo but it undergoes considerable enlargement before germination.

   In the Rosaceous plants, the embryo does not show any visible morphological or histological change during the after-ripening period. It is assumed that these seeds undergo physiological maturation during the dormancy period. The physiological changes may include the disappearance of growth inhibitors, appearance of growth promoters etc.
4. *Combined Dormancy*: Combined dormancy is a condition in which two or more primary factors, such as seed coat dormancy and embryo dormancy, are present to the extent that each requires treatment to overcome. Some examples of combined dormancy are *Cercis canadensis* var. *mexicana, Rhus trilobata.* Seed coat scarification with acid for 10 to 40 minutes followed by moist stratification, is the recommended treatment to overcome dormancy.
5. *Double Dormancy*: Double dormancy is a condition in which there is dormancy in both the radicle and the epicotyl of the embryo, but each have different conditions to overcome the dormancy, e.g., Oaks.
6. *Secondary Dormancy*: Secondary dormancy results from action, treatment, or injury to seeds during collection, handling or sowing. Pine seeds can incur secondary dormancy if exposed to high temperatures and moisture at critical times. When stratified seeds are re-dried to

storage levels, they are often said to have incurred secondary dormancy. Germination can certainly be delayed under these conditions.

Methods for Overcoming Seed Dormancy

1. *For seed coat dormancy*: Treatments are designed to breach the seed coat, or other covering structures, and remove barriers to moisture uptake, gas exchange, swelling of the embryo, and radicle emergence. Methods used to overcome seed coat dormancy are collectively known as scarification treatments, and there are risks to seed viability inherent in all of them. In selecting a scarification treatment, the most gentle methods should be tested first; then increasingly severe treatments until the desired effect is obtained.

   Cold water soak: In some hard seeded species, the seed coats are not completely impermeable to water. Soaking such seeds in water at room temperature for 24 to 48 hours may be sufficient for full imbibitions and subsequent germination.

   Hot water soak: Similar to the cold water soak, seeds are put into very hot or boiling water and left there as the water cools for 12 to 24 hours. The hot water softens the seed coats or causes them to crack, and imbibitions occur as the water cools. The container used for this treatment should not be made of aluminium, as it may be toxic to the seeds. Numerous leguminous species can be treated in this manner.

   Hot wire: This technique requires a heated needle or an electric woodburning tool to burn small holes through seed coats. A belt driven burner that scarifies seeds electrically shows promise for treatment of larger lots.

   Acid treatment: Treatment with concentrated sulphuric acid is the method of choice for many species. Seeds should be in contact with the acid for 15 to 60 minutes, depending on species or individual seed lot, and washed thoroughly in running water afterwards to remove any acid that remains on the seed coats, e.g., *Gymnocladus dioicus, Ceanothus velutinus*.

   Mechanical treatment: Mechanical scarification is used extensively for large lots of seeds. Mechanical scarification can be done by rubbing seeds between two pieces of sandpaper. Seeds may also be mixed with coarse sand, small rocks and shaken vigorously in a jar. Care must be taken not to injure the embryo. A mechanical device has also been developed to crack peach seed coats. For small samples, seed coats can be scarified by hand with knives, files, dippers, sandpaper etc.

2. *For internal dormancy*: Treatments to overcome internal dormancy are expected to bring about physiological changes within the embryo that will enhance rapid germination.

   Stratification (pre-chilling): The usual procedure for stratification is to refrigerate fully imbibed seeds at 1 to 5°C for 1 to 6 months. This procedure stimulates the natural winter conditions of temperate seeds that are lying on forest floor. During stratification (a) enzyme systems are activated, (b) stored foods are changed to soluble forms; (c) the inhibitor/promoter balances change.

   The optimum length of the stratification period varies greatly among species and among different seed lots of the same species. One benefit of stratification for nurseries is an increased uniformity of emergence. The low temperature used in stratification inhibit germination of the seeds that are no longer dormant while the remaining seeds are undergoing the needed internal changes. When the seeds are finally sown in favourable temperatures, there is a flush of uniform germination and emergence, which is crucial to even seedling development.

Importance of Seed Dormancy

1. In the stratification method uniform germination and seedling emergence can occur.
2. Dormancy is a great advantage when one wants to store seeds, but a disadvantage when prompt germination is desire.
3. Plant enables the embryo to safely pass the unfavourable part of the year and germinate when the conditions are suitable. Hence it is vital mechanism that connects one generation with the next.
4. Seeds are exhibiting primary dormancy when they are dispersed from the parent plant in a dormant stage. Prevention of germination during seed development may be due to the endogenous ABA content of the seed. Absence of or insensibility to ABA during seed development, results in the production of viviparous, e.g., Maize viviparous (vp), tomato sitiens (sit), etc.

CHAPTER XIX

SENESCENCE AND ABSCISSION

SENESCENCE

Senescence may be simply defined as those changes which lead sooner or later to the death of an organism or some part of it. The term abscission is used to describe the processes involved in the shedding of plant structures, characterized by the degradation of cell walls at the point of weakening. Present evidence suggests that cells surrounding the fracture line produce and secrete cell wall degrading enzymes which hydrolyze the central region of the wall, allowing the cells to separate and fracture to occur. In contrast, the less common process of 'mechanical tearing' involves the generation of large forces which tear apart an inherently weak band of cells. The cells along the fracture line thus play a passive role in this process, the flaking of bark providing a familiar example.

Distribution of Senescence Processes in the Life Cycle of Plants

Senescence processes are encountered in all plants and at all stages of the life cycle. Fig. 19.1 is a schematic diagram of the life cycle of a flowering plant in which senescence processes associated with successive stages of life cycle are listed.

A particular feature of this scheme is that the events which are listed are integral components of the developmental programme of the plant, with the notable exception of those that occur within the seeds.

Metabolic Changes Associated with Senescence

Chlorophyll degradation: One of the cardinal events by which senescence in leaves is generally recognized is the loss of chlorophyll, revealed in the change from green to yellow coloration.

Free radical damage in absence of repair process

1. Exhaustion of reserves
2. Failure of DNA repair
3. Loss of capacity for protein synthesis

1. Senescence of cotyledons
2. Senescence in differentiation of tracheary elements, sieve tubes, and some specialized cells
3. Senescence in turnover of root hairs and root cap cells.

Dry seed I → Imbibed seed II → Seedling III → Vegetative plant IV → Flowering plant V → Fruiting plant VI

1. Senescence of whole plant (monocarpism)
2. Senescence of aerial shoots (herbaceous perennials)
3. Senescence and dehiscence of dry fruits
4. Senescence (ripening) and abscission of fleshy Fruits

1. Continuation of many senescence processes present in vegetative plant
2. Senescence and abscission of floral parts

1. Senescence in cell differentiation and cell turnover as in seedlings
2. Senescence in 'self pruning' of branchlets in formation of tree canopies
3. Sequential leaf senescence
4. Seasonal, synchronous leaf senescence
5. Senescence in development of thorns and spines
6. Abscission process

Fig. 19.1: Summary of the distribution of senescence processes in the life cycle of seed plants. Most of the phenomena listed, except those which occur within the seed, are integral part of the development programme of the plant (after Sexton and Woolhouse, 1984).

The synthesis and degradation (turnover) of cellular constituents: In many species there is a progressive decline in the amount of proteins and RNA in the leaves in the course of their maturation and senescence. The loss of these essential cellular constituents is considered a potential causal factor in senescence.

Influence of Hormones on Senescence

The experimental approach most frequently used to study the effects of growth substances on leaf senescence has been to place excised leaves with their cut petioles in solutions containing the growth substances, or to float excised leaf disks on the appropriate solutions and observe the rate of senescence, usually in terms of loss of chlorophyll and protein. The results obtained varied greatly between species. In excised leaves of *Xanthium* and many other species, cytokinins delay senescence. Indole acetic acid is needed to delay senescence of leaves of *Prunus* species. Gibberellins will delay the senescence of excised leaves of *Taraxacum officinale* but cytokinins are without effect. In *Rumex* species and *Tropaeolum majus*, cytokinins or gibberellins will delay the senescence of excised leaves over similar ranges of concentration. The reason for this diversity of responses is not known. In the case of *Tropaeolum* it was found that the gibberellin content of the leaves decreased as they senesced.

Exogenously applied abscisic acid (ABA) accelerates senescence in leaves of many species. Treatment of leaves with ethylene will promote senescence. Auxin treatment of many plant tissues leads to an enhanced production of ethylene.

ABSCISSION

The more conspicuous examples of abscission such as the shedding of fruits, leaves, bud scales, floral structures and branchlets will be familier, but it is important to realize that virtually any aerial part of a plant can be shed in this way. Structures may range in size from the complete shoot system of tumble weeds down to the hairs shed from developing leaves. Fracture is caused by wall breakdown in 'separation layers'. Plants do not have the ability to produce separation layers anywhere; they seem generally limited to specific locations called abscission zones and are characteristic of a given species.

The active separation of a leaf from a branch, without injury to the branch, is called leaf abscission. Leaves of gymnosperms and woody dicotyledons are usually shed as a result of changes that take place in the tissues of the leaf base prior to leaf death. In the base of mature deciduous leaves a narrow zone, the so called abscission zone or region can be seen. The vascular system in this region is usually concentrated in the centre and the sclerenchyma and collenchyma are less well developed or absent. Two layers are discernible in the abscission zone, an abscission or separation layer, through which the detachment occurs and the protective layer, which at the leaf fall protects from desiccation and invasion by parasites to the exposed surface. In most leaves, flowers, fruits and some stems the preparation of the abscission zone occurs during ontogeny, but may also occur a short time before shedding of the organ in response to conditions causing abscission.

Several histological processes were observed during abscission, not all of them in the same species. Cell division may precede the actual separation, the new cell walls being specifically affected by the subsequent degradation phenomena.

In the abscission zone lignification of parenchyma cells, tylose formation in the tracheary elements, and callose deposition in sieve elements was also observed. The final event in abscission involves the enzymatic degradation of cell walls. This starts with the removal of calcium and pectin from the middle lamellae and continues with the hydrolysis of the cellulose walls. At the end the sclerified tracheary elements rupture. The middle lamellae between the cells of the vascular tissue also disintegrate. Xylem vessels and interstitial parenchyma of vascular bundles may also become weakened by hydrolysis of cell-wall material prior to separation. Golgi vesicles may be involved in secretion of abscission-specific enzymes. In most monocotyledons and herbaceous dicotyledons no dissolution takes place and abscission is apparently effected by physical stresses.

The protective layer, or cicatrice, is formed through deposition of protective substances such as suberin and wound gum, in the cell walls and the intercellular spaces. In woody species the protective layer is sooner or later replaced by periderm that develops beneath the protective layer in continuity with the periderm elsewhere in the stem.

It has been shown that auxin applied to leaf blades inhibits the abscission of the leaf. Ethylene, on the other hand, is regarded as reponsible for the induction of the enzymes active in cell-wall degradation. Peroxidase brings about an increased synthesis of ethylene, and phosphatase is associated with the aging phenomena.

CHAPTER XX

ANATOMY IN RELATION TO TAXONOMY

The first attempt to use anatomy in plant classification was made by Bureau (1864). He used the anatomical characters for delimitation of taxa of various levels in the family Bignoniaceae. Solereder (1899) in *Systematische Anatomie der Dicotyledonae* emphasized the importance of xylem structure in plant classification. Bailey (1951) and Metcalf and Chalk (1950) have listed numerous anatomical features that can be made use in taxonomy. The *Anatomy of Dicotyledons* (Metcalfe and Chalk, 1950) and *Anatomy of Monocotyledons* (Metcalfe, 1960 onwards) are some of the books in this direction.

EPIDERMAL CHARACTERS

Stomata

Leaf characters especially epidermal and stomatal characters are very much useful in taxonomy. The importance of epidermal characters was emphasized by Stace (1965) while the micromorphology (study of morphological characters with scanning electron microscope) has been emphasized by Dehgan (1980) and Barthlott (1981).

Van Cotthem (1970) divided the stomata into 8 types. They are:

Anomocytic (Ranunculaceous type)

Stoma surrounded by a limited number of cells that are indistinguishable in size, shape or form from those of remainder of epidermis. e.g., Ranunculaceae, Malvaceae, Papaveraceae, *Tridax procumbens, Boerhavia diffusa, Cleome* spp.

Anisocytic (Cruciferous type)

Stoma surrounded by three cells of which one is distinctly smaller than the other two, e.g., Cruciferae, Solanaceae.

Paracytic (Rubiaceous type)

Stoma accompanied on either side by one or more subsidiary cells parallel to the long axis of the pore and guard cells, e.g., Rubiaceae, Magnoliaceae.

Diacytic (Caryophyllaceous type)

Stoma enclosed by a pair of subsidiary cells whose common wall is at right angles to the guard cells, e.g., Caryophyllaceae, Acanthaceae.

Tetracytic

Four subsidiary cells are present, two lateral and terminal, e.g., most families of Monocotyledons.

Actinocytic

Stoma surrounded by a circle of radially elongate subsidiary cells, e.g., Commelinaceae, Musaceae.

Cyclocytic

Stoma surrounded by four or more subsidiary cells which form a ring around each stomata, e.g., tribe Laguncularieae of the family Combretaceae.

Hexacytic

Stoma accompanied by six subsidiary cells consisting of two lateral pairs parallel to the long axis of the pore and two polar (terminal) cells; the second lateral pair as long as the stomatal complex.

Besides the above, several other characters can be used in taxonomy. Presence of stomata on the upper side (epistomatic) or lower side (hypostomatic) or on both sides (amphistomatic) are also useful in taxonomy. For example, the presence or absence of stomata on the upper surface of a leaf is a good diagnostic feature in the separation of *Alnus subcordata* and *A. orientalis* of Betulaceae, which are otherwise difficult to distinguish. In the former epistomatic condition prevails while in the latter, the hypostomatic. Similarly, shape of the guard cells are also useful in plant classification. For example, stoma with bean-shaped guard cells are met with Dicotyledons while stoma with dumb-bell-shaped guard cells are seen in Poaceae. In Cyperaceae rectangular guard cells are seen.

Development of stomata has been divided by Pant (1965) into three types.

Perigenous

Stoma in which the subsidiary cells do not have a common origin with the guard cells, but are formed by cells lying around the meristemoid that divides to form the guard cells.

Mesogenous

Stoma in which the subsidiary cells are produced from the same meristemoid (initial) as the guard cells.

Mesoperigenous

Stoma in which at least one of the subsidiary cells has a common origin with the guard cells, but others do not.

Development of stomata is useful in plant classification. Ramayya (1977) has emphasized the importance of stomatal characters in taxonomy. Ramayya and Rajagopal (1968, 1971) gave a key to genera of Portulacaceae and Aizoaceae based on epidermal characters.

In the family Acanthaceae the stomata are diacytic, whereas in the closely related Scrophulariaceae the stomata are anomocytic. In the family Combretaceae the stomata are paracytic in the subfamily Strephonematoideae and anomocytic in the subfamily Combretoideae except the tribe Lagunculaireae where they are cyclocytic. Stomatal characters are not, however always so reliable and the different types have obviously each originated on many different occasions and they often exist together on one plant. In the genera *Streptocarpus* and *Saintpaulia* of the Gesneriaceae; for example, the cotyledons bear anomocytic stomata and the mature organs anisocytic stomata (Saharsrabudhe and Stace, 1974). In *Lippia nodiflora* of Verbenaceae anomocytic, anisocytic, diacytic and paracytic stomata occur on the same leaf (Pant and Kidwai, 1964). In *Alternanathera pungens* of family Amaranthaceae also all the above four types of stomata are found.

Inamdar (1968) who studied the stomata of *Nyctanthus* expressed the opinion that this genus should be kept in Oleaceae. In *Nyctanthus* and Oleaceae the stomata are tetra-mesoperigenous (sometimes aperigenous), paracytic while Verbenaceae shows dia-mesogenous stomata. So Inamdar (1968) said the *Nyctanthus* should be kept in Oleaceae instead of Verbenaceae.

Bremekamp (1953) said that the genus *Elytraria* should be removed from Acanthaceae and should be included in Scrophulariaceae. Paliwal (1966) studied the ontogeny of stomata to ascertain this. He found that in *Elytraria* and Acanthaceae the stomata are diacytic and development is syndetocheilic, but in Scrophulariaceae stomata are anomocytic and development is haplocheilic. Hence Paliwal (1966) retained the genus *Elytraria* in the family Acanthaceae.

Trichomes

Trichomes are epidermal outgrowths. They exhibit great variation in structure and function and thus are of much taxonomic value. Glandular and non-glandular hairs, papillae, setae and prickles are all considered as hairs. These trichomes are useful in plant classification and understanding the relationship between groups. Schulz (1936) used the type of hair as the major criterion in the subdivision of the family Cruciferae into tribes and genera. Ramayya (1962, 1969) studied the trichomes of Compositae and gave key to the genera based on trichome characters. Based on trichome characters, he considered Heliantheae as the primitive tribe. Rollins (1944, 1945, 1946, 1949) studied the trichomes of *Parthenium argentatum* and *P. incanum* and their natural hybrids. He has shown that in *P. argentatum* the trichomes are T-shaped while in *P. incanum* they are whip-like with a long thread. Intermediate types of trichomes are observed in the hybrids between these species. In the family Combretaceae, Stace (1965, 1969 a, b, 1973) found that trichome anatomy is of immense significance in classification at all levels, from the circumscription of the family down to the separation of species and even varieties. In particular, it has led to an improved tribal classification within the family and an improved subgeneric and sectional classification within the largest genus *Combretum*.

Epidermal Cells

The epidermal cells, other than those modified by their relationship to trichomes, stomata, the venation system or other special structures, provide many characters of taxonomic value. These include their size, shape, orientation, anticlinal wall undulation and periclinal wall curvature. The epidermal cell modification occurs also on the leaf margin and when silicified, calcified, suberized, gelatinized or containing various types of crystals. Silicified and suberized epidermal cells are of major taxonomic importance in the Poaceae.

Other epidermal features are also useful in plant classification. The genus *Morus* is divided into species based on idioblasts in epidermis. Tomlinson (1956, 1959 a, b) and Metcalfe (1968) said that silica bodies in the leaves are useful in classification of the families Zingiberaceae, Musaceae, Arecaceae and Rosaceae up to genera and even species.

Internal Structure

Carlquist (1961) said that leaves in Angiosperms show great diversity and this character is useful in classification. The nature of epidermis, its thickness, type of mesophyll, sclerenchyma patterns, sclereids, crystals, venation patterns, etc., may be useful in classification. The importance of foliar anatomy in the systematics of grasses was realized quite early by Duval Jouve (1875). According to Avdulov (1931) Poaceae have two types of internal structure in leaves. In the Festucoid (Pooideae) grasses the mestome around the stele is layered and mesophyll cells are straight while in Panicoid grasses mestome may or may not be present and the mesophyll cells are arranged radially.

Similarly, Ayensu (1974) found that sclerenchyma patterns in the Velloziaceae can be used to separate the genera *Vellozia* and *Barbasenia*. The genera in Cyperaceae can be segregated based on internal structure of the leaves (Govindarajulu, 1969).

Crystals, especially calcium oxalate crystals, are found to be useful in the classification of Angiosperms. Foliar sclereids have, of late, been demonstrated to be of taxonomic importance in many taxa.

Laticifers are cells or a series of fused cells containing latex and are common features of many succulent plants and other plants of arid regions. They vary widely in their structure and latex in their composition. Their presence or absence and when present, their structure has been of some taxonomic value. In Aroideae (De Bary, 1884), for example, certain species lack laticifers or any related structures while others have longitudinal rows of elongated, cylindrical sac-like cells. Some authors consider that Fumariaceae, which lack laticifers (Sperlich, 1939), should be considered distinct from the related family Papaveraceae having a well developed latex system. In the former family laticifers are replaced by idioblasts. It is one of the reasons for considering these two families as distinct from the rest of the families in Engler's Rhoedales.

The taxonomic usefulness of petiole vascularization patterns has become increasingly apparent in recent years (Radford *et al.*, 1974). Sinnot (1914) gave a classification of nodal types in megaphyllous plants. In general, there are three major types of node — unilacunar, trilacunar and multilacunar. Trilacunar nodes occur in the majority of Dicotyledons, multilacunar nodes in the primitive orders such as Magnoliales, Piperales, Trochodendrales and advanced orders such as Umbellales and Asterales and unilacunar nodes in Laurales, Caryophyllales, Ericales, Diapensiales, Ebenales, Primulales, Myrtales and a majority of families of Asteridae (Paliwal and Anand, 1978).

Stem anatomy has been studied in the herbaceous members where it has proved to be of some diagnostic value. In the stem of *Ranunculus repens* there is relatively little strengthening tissue and is correlated with the creeping habit of the plant, while in the closely related, species, *R. acris*, interfascicular tissue is sclerified and therefore an erect habit is possible.

Stace (1970) has shown that in the subgenus *Genuini* of *Juncus*, anatomical characters of stem and foliar bract can be used to distinguish most British species. Further, it is also possible to identify parents of several hybrids on anatomical grounds.

There is considerable variety in the anatomy of stems of *Dioscorea* and can be successfully employed in the delimitation of species that are not easily separable on exomorphic grounds. Ayensu (1970) found anatomy useful in separating *Dioscorea rotunda* and *D. cayennensis*, which are closely related and difficult to distinguish.

The classical case of anatomy in relation to taxonomy is the distribution of vascular bundles. In Dicotyledons vascular bundles are arranged in a ring while in Monocotyledons vascular bundles are scattered. Bentham and Hooker (1862–83) included Basellaceae as a subfamily in Chenopodiaceae. Anatomically Basellaceae are characterized by a single ring of vascular strands with intraxylary phloem in the larger bundles. On the other hand, Chenopodiaceae are characterised by the development of polycyclic, anomalous secondary vascular bundles.

Bentham and Hooker (1873–1876) split the Gentianaceae into two subfamilies — Menyanthoideae and Gentianoideae. But Hutchinson (1969), Cronquist (1981) and Takhtajan (1980) raised them to independent families. Vascular bundles in Menyanthaceae are collateral while Gentianaceae show bicollateral vascular bundles.

The most important anatomical features that have been used in taxonomy and phylogeny are those of the secondary wood. This is because of their conservative nature. Wood anatomy can be used at all taxonomic levels. At and above the family level, it has been successfully used along with other lines of evidence to establish the systematic position of the primitive vesselless Angiospermic families Winteraceae (all seven genera-*Drimys, Pseudowintera, Bubbia, Belliolum, Exospermum, Zygogynum* and *Tetrathalamus*), Trochodendraceae, Tetracentraceae, Amborellaceae and Chloranthaceae. Similarly it has been agreed by all phylogenists that the Englerian group of primitive Angiosperms, namely Amentiferae can not be considered primitive since they have specialized wood (Tippo, 1938). Anatomical evidence also supports the case for considering the remarkable genera *Paeonia* and *Austrobaileya* as the type of independent families (Bailey and Swamy, 1949).

At and below the generic level, the most useful contribution of wood anatomy is to provide evidence for placing of taxa of uncertain affinity. In *Quercus* (Fagaceae), different subgenera and sections can be delimited with the help of anatomical evidence. The placement of the genus *Myristica* and the related genera close to Lauraceae is supported by the evidence of wood structure (Eames, 1961), while on the same basis Myristicaceae are considered unrelated to Annonaceae and Eupomaticeae. Similarly, placing of Calycanthaceae in or near Rosales or Myrtales is not supported by wood structure.

Sieve cells also provide taxonomical evidence. For example, unspecialized sieve cells occur in lower vascular plants — Pteridophytes and Gymnosperms. Only one Angiosperm, *Austrobaileya scandens*, shows unspecialized sieve cells in phloem.

Floral Anatomy and Taxonomy

Floral anatomy has a wide application in determination of status of high ranking groups such as genera and species. On the basis of similarity of floral anatomical pattern affinities of families Papaveraceae, Capparaceae, Brassicaceae and Magnoliaceae could be established. These families have a common parietal placentation and inverted placental bundles with regard to floral axis on inner side of secondary marginal strands with normal orientation. This fact also supports placement of Moringaceae in Rhoedales. Puri (1947, 1949) suggested affinities between Passifloraceae, Moringaceae and Cucurbitaceae on floral anatomical basis.

Burtt (1965) opined that *Cyrtandromoea* has no close affinity with other members of Gesneriaceae and advocated its transfer to Scrophulariaceae with which it exhibits more resemblance. Singh and Jain (1978) have supported this treatment on the basis of floral anatomy. Kale and Pai (1979) have supported the removal of the genus *Trichopus*, on the basis of floral anatomy, from the family Dioscoreaceae of Bentham and Hooker and erection of a unigeneric family Trichopodiaceae as has previously been suggested on anatomical evidence by Ayensu (1966, 1972).

Fruit Wall and Seed Coat Anatomy

Fruit wall and seed coat anatomy have also come to the help of taxonomists in many cases. The distinction between the two genera *Ammannia* and *Rotala* (Lythraceae) has been puzzling taxonomists for quite sometime. The diagnostic feature was in the fruit-irregularly rupturing fruits in *Ammannia* and valvular capsules in *Rotala*. But this difference was very difficult to appreciate, especially in herbarium specimens. Van Leemjen (1971) found the fruit-wall character to distinguish the two taxa. The orientation of the cells of the two main layers of pericarp are different in *Ammannia* the cells of both layers are very similar, whereas in *Rotala* the cells of the inner layer are linear, strongly lignified and transverse to those of the outer.

Two different views were expressed regarding the systematic position of the Asteraceous genus *Lagascea*. Cassini (1815), Lessing (1832), DeCandolle (1836) and Merxmuller (1954) included the genus *Lagascea* in the tribe Vernonieae. Bentham (1873), Hoffmann (1894) and others place *Lagascea* in the tribe Heliantheae. Stuessy (1976) on the basis of morphological and chromosomal studies, supports the assignment of the genus *Lagascea* to Heliantheae. Pullaiah (1981) on the basis of pericarp structure favours retention of *Lagascea* in the tribe Heliantheae. In Heliantheae three different zones are distinguished in the pericarp. Between the middle and outer zones characteristic schizogenous cavities develop which get filled with a brown resinous and hardening substance. Such schizogenous cavities are absent in Vernonieae

Seed coat morphology has yielded valuable taxonomic informations. Seed coat anatomy has also been used for systematic purposes.

Ultrastructure in relation to Taxonomy

Morphology and taxonomy have benefitted greatly in recent years from information obtained by using advanced technology in microscopy, particularly Scanning Electron Microscopy (SEM) and Transmission Election Microscopy (TEM).

However, only a few ultrastructural characters are of wide application in plant classification. 'Dilated Cisternae' (DC) in the endoplasmic reticulum were first noted and described by Bonnet and Newcomb (1965) in the root cells of *Raphanus sativus* (radish) and later by Favali and Gerola (1968) in the phloem parenchyma of foliar veins in *Brassica chinensis*. Jorgensen *et al.* (1977) detected them in foliar veins in *Capparis cynophallophora*. Iverson and Flood (1970) found that such structures are common occurrence in the Cruciferae. According to Behnke (1977) these organelles are typical of Brassicaceae and Capparaceae.

Behnke (1972, 1977, 1981) in a series of observations since about two decades ago, has demonstrated ultrastructural details of sieve element plastids, which can be used in plant taxonomy. Behnke studied more than 2500 species from 380 families with regard to this character. There are broadly two types of such plastid elements, one accumulating starch (S-type) and the other accumulating protein (P-type). There are, however, certain groups of plants-Rafflesiaceae, Crassulaceae and some species of Ulmaceae, Moraceae and Urticaceae in which the plastids have neither starch nor protein accumulations, and they are called S_0-type. S-type sieve element plastids with starch accumulation in grains of different sizes are the commonest type in flowering plants. About 65 percent of these have such plastids in the sieve tube elements (Behnke, 1977). The subclasses Ranunculidae, Hamamelididae and Dilleniidae have without exception S-type sieve element plastids, while in the Rosidae a great majority of the orders have this type. Half the members of the Magnoliidae have S-type and half P-type elements. In Caryophyllidae majority of the members show P-type sieve element plastids.

P-type plastids are further differentiated on the number and shape of the crystalloids and the nature of the filaments surrounding them.

(i) P I subtype: The plastids contain single crystalloids of different sizes and shapes and/or irregularly arranged filaments. P I subtype sieve element plastids are present in Magnoliales, Laurales and Aristolochiales.
(ii) P II subtype: The plastids contain several cuneate crystalloids oriented towards the centre of the plastid. P II subtype sieve element plastids are present in Monocotyledons.
(iii) P III subtype: The plastids contain a ring-shaped bundle of filaments. These plastids are seen only in Centrospermae.
(iv) P IV subtype: The plastids contain a few polygonal crystalloids of various sizes. This type of plastids are seen only in Fabales.
(v) P V subtype: The plastids contain many crystalloids of different sizes and shapes. It is seen in the order Ericales and family Rhizophoraceae.
(vi) P VI subtype: The plastid contains a single circular crystalloid. It is found in family Buxaceae.

Behnke (1974) investigated the order Rhamnales and found that the family Rhamanaceae had S-type while Vitaceae and Leeaceae had P-type sieve element plastids. The heterogeneity of the Magnoliidae as to its plastid types is attributed to its basal position in the evolution of flowering plants (Behnke, 1975). P-type are considered to be ancestral and S-type has been thought to be derived by loss of protein. P I subtype is held to be the most primitive among P-subtypes of flowering plants.

The best use of plastids has been made in elucidating the interrelationships and circumscribing the order Caryophyllales (Centrospermae) (Behnke, 1976). All the families kept in this order show P III subtype sieve element plastids. These families are: Aizoaceae, Nyctaginaceae, Phytolaccaceae, Portulacaceae, Caryophyllaceae and Molluginaceae. In earlier classification other families like Polygonaceae, Vivianiaceae, Bataceae, Gyrostemonaceae, Plumbaginaceae, Fouquieriaceae, Frankeniaceae, Rhabdodendraceae and Theligoniaceae have been kept in Centrospermae but in recent classifications these have been removed from Centrospermae because they show S-type sieve element plastids.

CHAPTER XXI

PROGRAMMED CELL DEATH

Krishnamurthy *et al.* (2000) in an excellent review have given details of Programmed cell death in plants.

Programmed cell death (PCD), a usage coined by Lockshin (Lockshin and Williams, 1965), is a physiological process that leads to the selective elimination of unwanted cells in multicellular animals. PCD is characterized by the invariable involvement of the dying cells' own machinery as an 'executioner', i.e., it is a case of cell suicide. It is reported to play a major role in controlling cell number and turnover especially in the skin, gut and reproductive tract, tissue homeostasis and specialization, tissue sculpturing and pattern formation, oncogenesis and other disease incidents, defence reactions, and above all, in maintaining the overall integrity of the animal. PCD is distinct from necrosis. While the former is an energy-dependent physiological process that is genetically programmed with the involvement of regulatory genes, stimulatory events, signaling pathways and morphologically expressed distinctive features, the latter is a non-physiological process dissociated from developmental and morphogenetic events and involving cell swelling, lysis and leakage of cell contents with no obvious changes in DNA (i.e., no chromatin condensation and DNA fragmentation).

PCD has also been reported in plants, although plant PCD studies have been initiated with greater vigour only in very recent years. PCD is as important in plants as in animals and that it can occur on a local or systematic scale.

CATEGORIES OF CELLS THAT UNDERGO DEATH

A number of categories of cells can be recognized within those that undergo death in plants.

Category (i): Cells that have served their functions — A number of plant cell types belong to this category: root cap cells, cells undergoing senescence, cells involved in abscission of plant parts, vegetative cell of pollen, synergids, antipodals, anther tapetal and wall cells, transmitting cells of stigma, style and the ovary wall, endosperm cells, including haustoria, aleurone cells of grass endosperm, suspensor cells of developing embryo, etc. The cells of this category lose their earlier identity, morphology and functions and their remnants or corpses may persist. Available evidence strongly indicates that these cells undergo a physiological PCD; cells at the line of abscission may undergo PCD while the other cells of the organ may die indirectly because of not receiving nutrients, water or growth regulators.

Category (ii): Cells that are unwanted from the beginning — The stamen and carpel primordial cells, respectively in female and male flowers, non-functional megaspores in monosporic and bisporic categories, non-functional microspores in members of Cyperaceae and Epacridaceae, non-functional cells formed during embryoid initiation in cultured cells, etc. are examples of plant cells belonging to this category.

Category (iii): Cells that die during redifferentiation into specialized cell types — Xylem tracheary elements, cells of certain trichomes, thorns and spines, sclerenchyma (sclereids and fibres), cork cells, etc. are some of the plant cells belonging to this category. The most important characteristic of these cells is that they all start functioning only after their death. The other characteristic feature of these cells is that significant changes (morphological and chemical) take place in their cell walls. In other words, cell death leads to terminal differentiation in this category of cells.

A sub-category under this includes those cells which undergo 'partial'(?) death during differentiation. The phloem sieve elements belong to this category. (Nuclei persist in mature sieve elements, albeit with a lot of structural changes and loss of transcriptional activity in species of *Selaginella* and *Isoetes* and in some ferns.) Here the nucleus is lost, but not the cytoplasm and its integrity (?). The functional life span of the cell is, however, fixed; a sieve tube element functions from a few days to a few years depending upon the plant.

Category (iv): Cells that are subjected to hypersensitive response — Cells subjected to hypersensitive response (HR) due to pathogenesis and other environmental and abiotic stresses such as osmotic, oxidative (H_2O_2 and salicylic acid), temperature, salt, water, heavy metals, UV, nutrient deprivation, toxins, chemicals, etc. often undergo PCD. Here cell death is used as a defence against the stress or as a means of emanating signals for the other nearby cells to build up defence/immune reactions. PCD is reported to be employed in these instances not only to kill the host cells (to result in Hypersensitive resistance response — HR lesions) to prevent further infection, but also the invading pathogens. In many instances, cells experiencing lethal stress doses often react by activating their PCD mechanisms to commit suicide before they are killed. A critical survey of the available literature indicates that not all cells subjected to stress undergo PCD; there are cells which may undergo necrosis, especially if the dosage of stress is beyond a particular level.

Category (v): Cells that are damaged and unable to function further — In plants, cell death due to injury and other reasons may occur, but it is not known whether such deaths are programmed. It is often difficult to distinguish this category of plant cells from the previous category.

Category (vi): Cells that are present in wrong places — Cells that die to result in leaf lobe formation in lobed leaves and leaflets in compound leaves or to result in holes in the lamina of taxa like *Monstera* and some Croton varieties are typical examples of this category.

PROCESS ASSOCIATED WITH PCD IN PLANTS

PCD in animal system is reported to result in the disassembly of cells involving the condensation, shrinkage and fragmentation of the cytoplasm and nuclei into several sealed packets (often called the apoptotic bodies), which are then fagocytosed by the neighbouring cells or macrophages. Thus there are no remnants of cells corpses left. Nuclear fragmentation is preceded by chromatin condensation and marginalization in the nucleus. Fragmentation of DNA at the nucleosome linker sites then takes place. When all these events are combined to result in a distinct morphological expression then PCD is termed as apoptosis, a word coined by Kerr *et al.*, in the year 1972. In other words, apoptosis is a distinct form of PCD. However, there are others who consider apoptosis and PCD as one and the same. The apoptotic PCD involves a syndrome of mechanisms with effectors, adaptors, regulators and signals as well as the activity of endonucleases.

Although several plant systems have been studied to understand the mechanism of PCD, very little attention has been paid to the details of cytoplasmic changes, if any, during cell death. Condensation and shrinkage of the cytoplasm were reported in dying aleurone cells, root cap cells, sister cell of the embryonal initial, and HR lesion cells of *Arabidopsis*. In the embryoid sister cell, the break-down products of the cytoplasm are reported to form several small membrane-sealed packets. TEs of intact plants is reported to become lobed, condensed, shrunken and finally gets broken into small packets. A critical perusal of hundreds of published micrographs of dying plant cells belonging to megaspore dyads and tetrads, nucellus, endosperm, embryo, suspensor, anther tapetum, root cap, etc. indicates the occurrence of cytoplasmic condensation and shrinkage but not its breakage into bits; there are evidences of vacuolar autophagy in most cases which accounts for the elimination of the cytoplasm.

A different mechanism of programmed autolysis was first suggested by Nagl with reference to the death of embryo suspensor cells. The filiform leucoplasts of the suspensor cells are transformed into cytosome-like structures displaying high acid phosphatase activity. During transformation, polar swelling of the plastids takes place incorporating into it the invaginations of the degenerating cytoplasm to form spoon-like structures, which may later on become 'ball-like'. These structures were termed 'plasto-lysomes'. They form the characteristic elements of the lytic compartment. Plastolysomes include various organelles and cytoplasmic components, particularly ER, polysomes and lipid bodies. At a later stage, the stroma of these plastolysomes disappears completely and finally the bodies are transfomed into autophagic vacuoles surrounded by a single membrane; all the included cytoplasmic components also degenerate. Whether it be by vacuolar autophagy or by plastolysomy, the plant cells that died are always evident as corpses and these are invariably represented by the remnants of the cell wall.

Effect of Caspases

Animal PCD is due to a syndrome of well organized and executed mechanisms involving effectors, adaptors, regulators and signals. The chief effector molecules are caspases. No caspase has yet been cloned from plants. But there is a growing information on the family of caspase-like proteinase and their involvement in plant PCD. The expression of cysteine proteinases (CyPs) is reported during cell death in *Solanum melongna* (Xu, 1999). Several types of cysteine proteinases such as actinidin, oryzain, barley aleurain, etc. are known from the plant system. Phytepsin D, a barley vacuolar aspartate proteinase, is reported to be expressed very highly during autolysis and development of TEs and sieve elements. It is, however, interesting to note that these proteinases are expressed not only during cell death but also during other cell events. There is still only a speculation that proteinases reported in plant PCD are involved in a function analogous to that of cysteine proteases of the interleukin 1^2 converting enzyme (ICE) family reported in the animal system.

The substrates for almost all these plant proteinases are yet to be identified. It is quite possible that the plant proteinases are involved in the clearance of cytoplasmic and nuclear proteins from dying/already dead cells such as TEs, i.e., they may merely be doing a scavenging job, thereby helping the recycling of nitrogenous compounds. Moreover, the results of Xu and Chye (1999) in brinjal on the isolation and characterization of cDNA encoding CyPs and their expression indicate their coincidence with several rather than specific events in cell death in many organs/tissues such as ovules, xylem, placenta, receptacle, anther, epidermis, endothecium, nucellus, etc. Although CyPs of brinjal has 90–92% amino acid identity to the other CyPs from Solanaceae family (tobacco CyP 8 and tomato CyP 2), they show only 39% homology to cathepsin H, an animal protease. Caspase inhibitors, such as the chloromethylketone Ac-YVAD-CMK or aldehyde Ac-DEVD-CHO, are reported to completely abolish Hypersensitive resistant response (HR) cell death in tobacco. This is cited as an evidence for the presence of caspase-like plant proteases that participate in HR cell death. But it is not yet clear whether these inhibitors are targeting the proteases that are evolutionarily related to metazoan caspases.

There are reports of a gene implicated in 'caspase' production and function in plants. *SAG12*, an *Arabidopsis* gene, is reported to encode a putative cysteine protease involved in senescence. *HSR 203 J*, a tobacco gene, is activated during HR specifically, while *HINI*, a tobacco gene is activated during HR as well as lately in senescence, but the last two genes are reported to be activated only after the switching of 'cysteine proteases'.

Adaptor molecules and Regulator molecules

Neither Adaptor molecules nor Regulator molecules have been reported in plant PCD.

Signal molecules

Oxidative stress, as a signal event, is shown to trigger cell death in plants; the involvement of O_2 in plant Hypersensitive Resistance (HR) responses and of H_2O_2 in cell death mediated by pathogens are reported. However, we do not have a clear idea about the exact role of these two ROS in cell death process in both animals and plants, although H_2O_2 is likely to induce DNA fragmentation and an influx of extracellular calcium. In plants H_2O_2 has been reported not only in cells infected with pathogens and subjected to stress but is also regularly produced when there is an activity of peroxidase and polyaminic oxidase, lignification and cross-linking of extensins in the cell wall. It is true that H_2O_2 is present not only in living cells but also in cells that undergo death due to wounding, pathogen entry and xylogenesis, but available evidence is not sufficient enough to unequivocally implicate ROS species in PCD of these cells; they are more likely to be involved in necrotic cell death which is found hand in hand with PCD in systems subjected to biotic and abiotic stresses, especially at a level more than threshold.

Like in animal PCD, in plant cells undergoing death an increase in cytosolic calcium is reported to be an important signaling event. The probable role of calcium is the activation of endonucleases in both PCDs. However in plants Zn^{2+}-dependent endonuclease is known in additon to Ca^{2+}-dependent endonuclease. The former is known in cells of wheat leaf undergoing senescence, while the latter is known in differentiating TEs, aleurone cells and in HR lesion cells.

Plant cells subjected to oxidative stress (i.e., treatment of H_2O_2) showed an increase in intracellular reduced glutathione. Plasmamembrane blebbing has not so far been reported in plant cells undergoing death, although PS exposure has been demonstrated through Annexin V binding assay in the protoplasts of tobacco subjected to abiotic stress. Whether this has any relation to the presence of the cell wall in the plant system is not clear.

Chromatin condensation, endonucleases and DNA fragmentation

The programmed death of aleurone cells in barley during germination displays several characteristics associated with apoptosis. TUNEL staining of aleurone cells that had been triggered to die indicated that nuclear DNA in the cells undergoes fragmentation starting near the embryo. The fragmentation proceeds outward in a time-dependent manner *in vivo* and under osmotic stress in isolated cells *in vitro*. The visualization of DNA fragmentation precedes visible signs of death. Abscisic acid inhibits the aleurone cell death and DNA fragmentation both *in vivo* and *in vitro*. Okadaic acid, a protein phosphatase inhibitor that suppresses apoptosis in some animal systems, also inhibits the death of aleurone cells.

The root cap consists of living parenchyma cells derived continuously from the apical meristem. As new cells are produced in the interior, those on the periphery of the root cap are shed in a very orderly manner, reminiscent of Kerr's original connection of the term apoptosis to cell death as a natural programmed process. Cap cells of fresh onion and tomato root tips, double stained by the TUNEL assays and a DNA stain, revealed that nuclear DNA in cells throughout the root sections remained intact until the cap cells reached the surface of the root. The region of TUNEL-positive cells was restricted to the outermost 1–3 cells of the root cap. The appearance of nuclear DNA changed from a compact circular form to elongated TUNEL-positive structures that eventually took on the form of numerous membrane-bound circular bodies that migrated to the periphery of the dying cell. The formation of these bodies appeared to be the final step reached before a cell is shed from the root cap. The morphology of PCD in aleurone and root cap cells confirmed the conservation of at least the terminal phases of the apoptic process in plants.

Stress proteins

One very interesting feature about the death of some types of plant cells is the active production of special categories of structural proteins rich in hydroxy-proline, proline, glycine, arabino galactan and occasionally threonine. These are actively synthesized in the endoplasmic reticulum packed in the golgi and transported to the cell wall where they become integral components. In sclerenchyma cells and xylem elements such proteins have been repeatedly reported during the process of cell death.

REFERENCES

Gilchrist, D.G. 1998. Programmed cell death in plant disease — The purpose and promise of cellular suicide. *Annu. Rev. Phytopathol.* **36**:393–414.

Krishnamurthy, K.V., R. Krishnaraj, R. Chozhavendan and F. Samuel Christopher. 2000. The programme of cell death in plants and animals — A comparison. *Curr. Sci.* **79**:1169–1181.

INDEX